S0-BLS-431

SCIENTIST'S GUIDE TO POSTER PRESENTATIONS

SCIENTIST'S GUIDE TO POSTER PRESENTATIONS

Peter J. Gosling, Ph.D.
Peter Gosling Associates
Staines, United Kingdom
(Formerly of the Department of Health, London, United Kingdom)

Kluwer Academic / Plenum Publishers
New York, Boston, Dordrecht, London, Moscow

Chemistry Library

Library of Congress Cataloging-in-Publication Data

Gosling, Peter J.
Scientist's guide to poster presentations / Peter J. Gosling.
p. cm.
Includes bibliographical references and index.
ISBN 0-306-46076-9
1. Poster presentations. 2. Science--Research--Posters.
I. Title.
Q179.94.G67 1999
808'.0665--dc21
99-35732
CIP

ISBN 0-306-46076-9

©1999 Kluwer Academic / Plenum Publishers
233 Spring Street, New York, N.Y. 10013

10 9 8 7 6 5 4 3 2

A C.I.P. record for this book is available from the Library of Congress.

All rights reserved

No part of this book may be reproduced, stored in a retrieval system, or transmitted in any form or by any means, electronic, mechanical, photocopying, microfilming, recording, or otherwise, without written permission from the Publisher

Printed in the United States of America

To my wife, Jacqueline

Q
179
.94
G67
1999
CHEM

"The two words 'information' and 'communication' are often used interchangeably, but they signify quite different things. 'Information' is giving out; 'communication' is getting through."

Sidney J. Harris

PREFACE

Scientific information is increasingly being communicated at both national and international scientific conferences in the form of poster presentations. A recent international conference, for example, which involved nearly 500 delegates from over 50 different countries, included information presented in 331 posters.

Recent experience has shown, however, that the presentational standard of such posters, even at the international level, varies immensely. Individuals presenting well-designed, eye-catching, and engaging posters are at a distinct advantage in promoting their scientific information. In doing so, they also promote themselves as credible scientists, as well as promoting the reputation of their establishments and countries of origin. However, producing a poster of high quality requires considerable planning and the acquisition of specific presentational skills.

This book provides detailed practical guidance on all aspects of presenting scientific information in the form of posters. It is assumed that the presenter has access, either at home, at work, or through libraries, to a computer or word processor, a color printer, and photocopiers. The book is intended to help scientists to gain poster presentational skills in a manner that enables adoption of an individualistic style of presenting information in a credible fashion. The book has relevance internationally and is primarily targeted at research workers, including postgraduate students and all scientific professionals who are required to present visual communication of scientific information. It is anticipated that the readership will include a wide range of scientific professionals, such as medical and veterinary microbiologists, public health scientists, marine biologists, chemists, and engineers.

CONTENTS

SCIENTIST'S GUIDE TO POSTER PRESENTATIONS

1

INTRODUCTION

Changing trends in the communication of scientific information at national and international scientific meetings has resulted in the emergence of the use of posters as a means of visually conveying scientific information. Poster presentations have rapidly become accepted and are now the major form of communication at many scientific meetings.

The advancement and dissemination of scientific knowledge depends on effective communication of research findings by scientists to the scientific community, including colleagues, collaborators and peers, and the wider population. Traditionally, this has been achieved by formal oral presentations at local, national, and international meetings, and by written papers in peer-reviewed journals. Such communications can be considered to be the building blocks on which a scientist's career is built. All scientists must therefore communicate their ideas, opinions, and research findings as widely and as frequently as opportunities present if they are to establish a reputation of good standing and advancement in their chosen profession. It is not surprising, therefore, that there is a high submission rate to present information at scientific meetings. However, lack of sufficient meeting rooms and of adequate time for spoken presentations have made an alternative form of presentation a necessity. Visual presentations in the form of posters have emerged to fill that need. First introduced into scientific meetings in Europe and the United States in the 1970s, they rapidly became accepted as a way to display large numbers of research efforts.

The poster session strategy adopted at meetings varies, but generally the poster will be on display for several hours to several days, and the authors will be present during a part of that time to discuss the subject with viewers. Depending on the meeting, the number of posters displayed at one time may range from a dozen to several hundred. The audience, by the nature of the venue, is always a group of people who are interested in the subject, albeit to varying degrees on any particular aspect. Presenting a poster is a good opportunity to build your reputation as a confident, knowledgeable, articulate scientist if you exhibit an attractive, informative display and maintain a professional demeanor as the author. This book provides guidance on the skills required to achieve these aims.

Although conceived out of necessity, poster presentations provide certain strong advantages in communicating with your audience. These displays of research findings combine visual and verbal information by using illustrations, the written text, and a spoken explanation by an author. The organization and presentation of a poster plays an important role in the facilitation of active communication. In this publication I draw on my own experiences to advise how best to maximize the impact of a poster presentation.

This book cannot be a blueprint for success. The best conceived graphic or textual element can look dreadful in the wrong position; you must always consider the overall design of the poster before preparing any individual segment. The emphasis of this book is on principles, because everything should be adapted to your own needs and display. It is impossible for anyone to tell you how to produce a scientific poster element by element, as every scientific meeting is different and no two posters are exactly alike. However, I have included hints and ideas that you may be able to adapt to produce a poster that effectively communicates your message and suits your own style.

All of the various aspects of poster design and construction are within the scope of a scientist, engineer, or other professional with average ability. Some of the tasks are difficult, for example, précising large amounts of data or pieces of text, whereas other aspects, such as the use of color, may be subjective. There is no disgrace in recruiting some assistance with these, however, and I would actively encourage you to seek the views, intuition, and ideas of colleagues and of friends.

The book is arranged in a sequential manner, leading the poster presenter from the initial considerations and planning to the final presentation at the scientific meeting. Chapters 1 through 4 are concerned with the purpose of posters: the pros and cons of poster presentations and how to take advantage of the opportunities they present. Chapters 5 through 9 cover aspects of what to include: how best to present text and data and use color in the overall format and design of the poster. Chapters 10 and 11 deal with actual construction: how to produce posters in a variety of styles and formats, as well as general advice on paper characteristics, drawing aids, adhesives, and cutting equipment. Chapters 12 through 14 are a guide to transporting the poster to the meeting and presenting it at the poster session, and should take some of the mystique out of creating the "right" image to the people who matter.

Producing a high quality, effective poster for display at a scientific meeting may seem a formidable task at first, but this need not be the case if tackled step by step. For the young or inexperienced scientist, I hope this book will give you that extra confidence and encourage you to make a start to contribute poster presentations to scientific meetings. For the experienced poster presenter, I hope by conveying my professional experience, thoughts, and ideas, you will be enthused to continue to produce and to improve on your poster presentations and will benefit more from the overall experience.

2

THE FUNCTION OF POSTER PRESENTATIONS — AND THE OPPORTUNITIES THEY OFFER

Many scientists who have experience in presenting information, both by traditional oral presentations and by poster displays, would agree that the latter is not always the easier option. Designing, constructing, and presenting a poster may require a considerable investment of time, effort, and money. However, the number of posters offered for presentation for the majority of scientific meetings continues to increase. What is it about poster presentations that make them such a desirable means of communicating scientific information?

In this chapter we consider the advantages and disadvantages of poster presentations with respect to effective communication of scientific information and for meeting organization. Additionally, we consider some of the many opportunities poster presentations offer to the individual presenter and reflect on the ways on which they may be capitalized.

2.1. THE FUNCTION OF POSTERS

The ultimate aim of a scientific poster is "to convey scientific information and views in a visual format, to an interested audience." Although this statement is true, however, it does not allude to the many close personal interactions such communications may impart. Delegates to scientific meetings are there primarily to keep apprised of the latest developments in their speciality, but they can also engage in personal interaction with friends, peers, and authorities who may include research fund holders. Presenting a poster may therefore be a gateway to meeting and communicating, face to face, with those with knowledge, influence, and resources while you are in the role of an active participant to the scientific proceedings.

From the conference organizers' perspective, poster presentations have a further purpose. They enable more researchers to present their ideas at the meeting, bringing with them additional scientific information that can be included in the subsequent proceedings publication. This ultimately increases the prestige of the meeting and has the benefit of promoting the reputation of the organizers and raising the profile of the scientific subject of the meeting.

2.2. POSTER PRESENTATIONS VERSUS TRADITIONAL ORAL PRESENTATIONS

Traditional oral presentations at most large scientific meetings leave much to be desired. They follow a set timetable, which may or may not be followed, that commits the delegates to a particular time if they wish to hear the talk, which typically last only 10 to 15 minutes. During the talk, effective communication between a speaker and his or her audience may be hindered by a host of conspiring factors. These largely relate to the venue, the presentational skills and experience of the speaker, and the thoroughness of the planning and preparation for the address. All too frequently the venue is either too large or acoustically unsuitable to permit coherent dialogue during the few minutes allotted for questions at the end of the talk. During this time it is not unusual to restrict the number of delegates wishing to ask questions to five or six chosen at random and if the program is not running according to schedule, the whole discussion component may be omitted altogether. Additional factors that conspire to hinder communication include the quality of the oral delivery and slide preparation, which is often so poor that the scientific message is indiscernible.

In contrast, poster sessions consist of many display boards shown simultaneously so viewers may either wander through all of the displays or go directly to the ones that interest them. Participants remain with the display for a set period, generally 1 to 2 hours, to expand on the material and answer questions. Many authors and delegates are more comfortable in the smaller groups than at a formal paper presentation.

2.2.1. Advantages of Posters

Poster presentations have a variety of advantages over traditional oral presentations. They offer distinct advantages to organizers of scientific meetings and delegates.

2.2.1a. Meeting Organization. In terms of meeting organization, poster presentations allow the following:

- many more papers to be scheduled than with oral presentations, for the same amount of time
- the use of some convention centers that may be able to provide large areas for display more easily than they can provide numerous meeting rooms and audiovisual equipment

2.2.1b. Communicating Scientific Information. For the delegate, posters may present an advantageous format for communicating scientific work, particularly if the content of the paper is highly "visual." Posters enable great graphic flexibility. The ease of creating color graphics, combining photographic and graphic techniques, computer graphics, and outputs from recording devices increases the range of graphic options available to the author. Even three-dimensional elements may be used.

2.2.1c. Dialogue with Viewers. Poster sessions enhance dialogue with those interested, providing the following:

- increased time for intensive discussion of paper results
- an opportunity for questions or comments relating to clarification or expansion of points made or techniques adopted
- an opportunity for informal contact with viewers
- an opportunity for mutually beneficial interaction between the presenter and the audience
- a convenient means of establishing personal contacts and collaboration

2.2.1d. Other Notable Advantages. Poster presentations offer a number of other notable advantages. In contrast to the audiences of oral presentations, a large proportion of the audience of any particular poster are more likely to be genuinely interested. Delegates tend to "sit through" oral presentations, possibly waiting for another speaker or simply sitting while planning their next activity. Poster presentations enable delegates to mull over the work, including the finer details that would not be available during an oral presentation. Data and graphics on the poster are available for as long as individual viewers wish to see them. The viewer may therefore focus on aspects that have the most personal interest; a poster allows the viewer to go back and review a figure or some text, or take notes at will.

When the meeting is over a poster may still have some use. Many authors mount them at their own institutions afterward to communicate recent research to departmental colleagues and to students. They may also serve as valuable visual aids for visiting scientists and for site visits.

2.2.2. Disadvantages of Posters

The emphasis on posters does however present some potential disadvantages, such as the following:

- Not all types of research results lend themselves to the poster format. For example, results from complex experiments involving large amounts of mathematical analyzes may be difficult to clearly present as a poster.
- Posters may require more time and expense to prepare than a presentation based on slides.
- Posters are more difficult to transport.

2.2.3. Outcome of the Appraisal

Overall, the case for the use of posters in communicating scientific information is compelling. Posters offer distinct advantages both for meeting arrangements and for communication efficiency. The rapid acceptance of poster sessions at scientific meetings and the growing number of contributions offered to meeting organizers exemplifies the advantages of this format.

2.3. OPPORTUNITIES PRESENTED BY POSTER PRESENTATIONS

Although poster displays offer you a chance to effectively convey your scientific message, poster sessions may confer a host of further opportunities to the astute delegate. For younger or inexperienced researchers in particular, dialogue at a poster session may provide important contacts and valuable feedback from scientists outside the immediate field. You may be able to obtain advice and opinions on specific aspects of your research or generally broaden your research perspective. It should enable you to form a network of contacts that may result in subsequent attraction of funding, collaborative research studies, and invitations to participate at future scientific meetings. Perhaps most important from a personal point of view, it is a period of high professional visibility with opportunities for you to promote yourselves, your working establishments, and the importance of your scientific research.

Some of the opportunities that may arise include the following:

- promotion of the authors
- potential to attract funding
- promotion of the authors' establishment
- promotion of the scientific information
- promotion of the authors' views
- acknowledgement by others working in the field
- establishing a network of contacts
- enhancing your professional reputation
- making a good impression on your peers

However, even for the most fortunate among us, simply accepting that such opportunities are likely to present themselves is probably not enough to reap a reward. The opportunities are there for the taking, but they have to be grasped. You will need to recognize as many of these opportunities as you can, and before setting off for the meeting, you should plan some achievable goals. Some possible goals and desired outcomes are shown in Table 2.1. You may find it helpful in planning your meeting strategy to

Table 2.1
Suggested Goals and Desirable Outcomes from Participating in Poster Sessions

Aim	Desired Outcome
To establish a wide range of professional contacts	Form and become part of a network of scientists working in collaboration.
To be noticed by others working in the field	Establish a good professional reputation, leading to collaborative work and possible acquisition of research funds.
To be noticed by the meeting organizers	Establish a reputation as a skilled presenter, with possible funding/invitations to participate in future meetings.
To discuss research findings with selected researchers	Obtain prepublication information or expert opinion on current research.
To promote your views	Dissemination of your particular scientific message.
To promote your employing establishment	Enhanced reputation of your employing establishment, leading to better funding and promotional prospects.

adapt this list of possible chances that could occur at a meeting in which you are participating and consider how you could capitalize on them. In considering your meeting strategy, study the list of speakers, as these will be particularly knowledgeable individuals and probably key players in their fields. Being early for their presentation or staying after they have finished may give you an opportunity to make contact, but meeting socially will almost certainly have a greater impact. Always plan to attend the meeting's social functions, mixers, conference dinner, and so forth, with this in mind. Be sure you have your business card or other means of contact details to offer. The list of participants, usually supplied with the meeting papers, is also a valuable source of identifying individuals with whom you would wish to make contact.

3

ESSENTIAL PREPARATION AND THINGS TO CONSIDER

3.1. PLANNING THE POSTER

Planning is the key to preparing a successful poster. With sufficient forethought, many potential pitfalls can be avoided, and your efforts can be focused on those aspects of production and presentation that really are important.

Careful planning will reduce the amount of time, effort, and expense required to produce a successful poster. I cannot emphasize enough the importance of this stage in poster production.

Those of you who are experienced or particularly organized and perceptive will have already recognized the value of routinely preparing visual elements, graphs, charts, photographs, and other forms of data from ongoing research studies. You may find that this practice is useful in consolidating and focusing your thoughts on a particular subject, and it has obvious advantages when it is time to present your findings. Similarly, periodically reviewing your work in writing, ideally with a view to publication in a peer-reviewed journal, will have the effect of focusing various aspects into manageable sections of text. If you have developed such a system, you will find gathering the variety of graphic and textual elements for your poster much easier. If not, you will have to start from scratch in drawing together the various elements for your display.

In general, a scientific poster must be clearly organized, contain an obvious central point, be easy to read from 1 or 2 m away, and be attractive and aesthetically pleasing. How you achieve this is largely a matter of personal preference and aptitude. It is the presenter's individualistic flair that provides the wealth of variety that makes poster displays both informative and entertaining. There are certain general aspects, however, that should be borne in mind when planning your poster.

3.1.1. Dimensions of the Poster

The dimensions of the poster will ultimately be restricted by the space available on the poster boards. This information should be supplied to you by the meeting organizers. *Take notice of these dimensions.* It is still not uncommon to see posters, even at prestigious international meetings, that are either too large for the poster boards supplied or so small that they appear completely insignificant. Whatever the dimensions of the board, you should not plan to have the body of your poster below waist level or above shoulder height. Any remaining space on the board above this portion can be used for the title, whereas any space below it may accommodate envelopes containing supplementary information sheets and so forth. The temptation to fill all available space on the board must be resisted. All too often, presenters fall into this trap and the result can verge on the ridiculous. I have seen viewers literally on their knees trying to read particular sections, but be aware that most will not go to this trouble and will simply move on to the next poster. Once you have decided on the most suitable poster dimensions, stick to them. You will then have the frame within which all the elements must be contained, and you will be in a position to start organizing the composition of your poster.

3.1.2. Composition of the Poster

You must consider what visual elements you need to include. These should be restricted to those considered absolutely necessary to visually convey your scientific message. It will also be necessary to give some thought to possible layouts. *It is essential to have a plan of the entire poster before attempting to produce any of the individual segments.* Too much or too little information will become apparent and editing can take place accordingly. This is best achieved by producing a rough sketch using blocks to represent sections of text and visual elements, an example of which is shown in Figure 3.1. Use this sketch to identify the number and relative sizes of the various sections of text, visual elements, headings, and so forth as well as the basic flow of information. It is probable that you will produce several revised

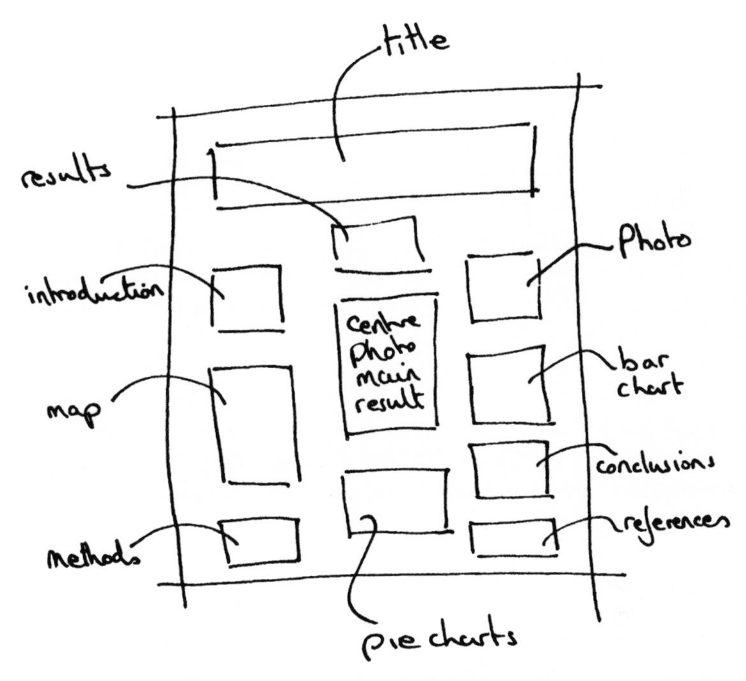

Figure 3.1. An example of a rough layout sketch. Rough sketches can be used to determine the basic format of the design and the flow of information. Many rough layout sketches may be required until you find a format that is especially appealing. It is an ideal opportunity to use your imagination to devise an attractive, eye-catching display.

and refined potential layouts before settling on the one you feel is the best. Once the rough layout of the poster has been prepared, it is necessary to move forward and produce a more detailed scale drawing of the layout, so the precise size and position of each element can be clearly ascertained. This may be conveniently drawn to scale on a sheet of graph paper. What could be included, the ways in which they could be presented, and preparation of scale drawings will be considered in more detail in Chapters 4 through 7.

3.1.3. Enlisting Assistance

You may need to recruit assistance in preparing certain elements of your poster that require expertise outside of your capability, or use of equipment that is not at your disposal. The need to utilize the services of photographers, computer operators, draughtsmen, or others, should be identified early in your planning stage. Firm arrangements should be secured to ensure your element will be prepared to your specifications and completed in time for your needs.

3.1.4. Cost

It is probable that few of those reading this book will have access to professional graphic designers and poster production facilities. If such professional services are commissioned, although it might be expected that the quality would be very good, the cost is likely to be prohibitive. It is more likely that the highest cost will result from producing suitable photographic elements. A smaller outlay will also be required for backing materials, adhesives, and so forth, but with careful planning these costs can be kept to a minimum. For most presenters the greatest cost will be in terms of the time that must be devoted to produce a successful poster. The amount of time can again be minimized with careful planning, but ultimately there is no way of avoiding this commitment.

3.1.5. Time

The time required to plan, prepare graphic elements, write and edit text, and construct the poster should not be underestimated. Sufficient time must also be allocated to write and edit the poster abstract. *It is never too early to begin.* I would recommend right from the beginning preparing a written timetable of events. Tasks should be listed and deadlines by which they should be achieved should be assigned to them. The deadlines must of course be feasible. They must also relate to the dates set for abstract receipt by the meeting organizers and, of course, the actual date that you are traveling to attend the meeting.

It is usually necessary to monitor closely the progress of various elements of the presentation that are outside of your immediate control; for example, photograph developing, printing, and enlarging. It is advisable to carry out these aspects as a planning priority, as it may take weeks to obtain the materials you require. Many scientists at this stage realize too late the advantage of compiling such materials concurrently while conducting their research.

A suggested schedule for the various preparatory steps is shown in Table 3.1. Although the tasks are listed in sequential order, the actual times you allot to them depend on your state of readiness and the deadlines imposed by the meeting organizers. The suggested time allowances may seen rather generous, but it is not until you come to carry out the tasks that you will realize that they are warranted. Some tasks identified in Table 3.1

Table 3.1
A Suggested Schedule for Some of the Preparatory Steps Leading to Poster Production

Task	Approximate Time to Achieve Task
Prepare timetable	2–3 weeks
Prepare outline plan	1 week
Prepare scale plan	2–3 weeks
Write poster abstract	3–4 weeks
Prepare photographs	4–6 weeks
Prepare graphic elements (graphs, charts, tables, etc.)	4–5 weeks
Prepare textual elements	3–4 weeks
Construct the poster	2–3 weeks
Review by several colleagues	2–3 weeks
Reconstruct and mount the poster	2–3 weeks

will be dependent on other factors listed. That is to say, they are sequential to a previous task; for example, you cannot complete the construction of the poster until the textual and graphic elements have been completed. To clearly ascertain the priority of tasks and to organize the most efficient timetable for their accomplishment I would recommend producing a bar chart (Gannt chart).

3.1.5a. Producing a Gannt Chart for Use in Poster Production. The preparation of a chart requires the following steps:

1. Assess the project and specify the basic approach to be used in production of the poster.
2. Break the production down into a reasonable number of tasks to be scheduled.
3. Estimate the time required to perform each task.
4. Place the tasks in sequence of time, taking into account the requirements that certain activities must be performed sequentially whereas others can be performed simultaneously. This should be shown on the chart.
5. Adjust the diagram to satisfy the time constraints imposed by the meeting deadlines.

These charts form a clear picture of the project and display planning and scheduling simultaneously. An example of a Gannt chart is shown in Figure 3.2. They can be used to show progress and to detect schedule slippage. They are most useful as they also specifically show the interrelationships, and hence the dependency relationships, among the tasks.

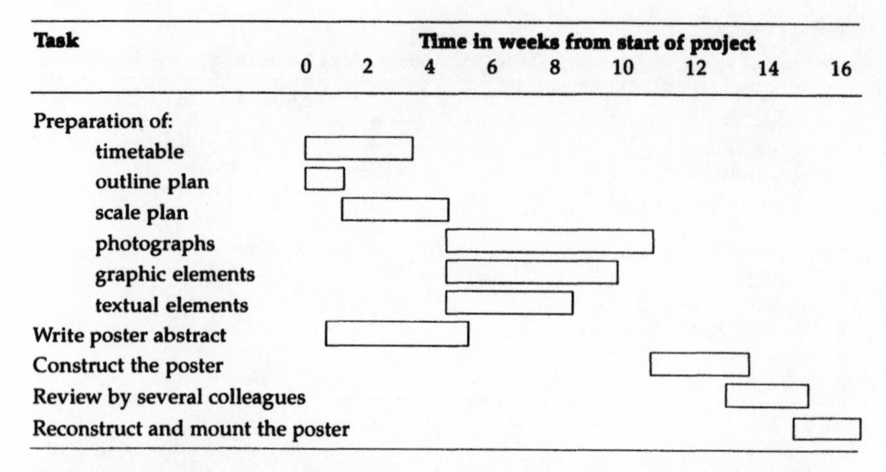

Figure 3.2. An example of a Gannt chart. This chart illustrates the suggested schedule for poster production, shown in Table 3.1. You should use your Gannt chart to monitor the progress of your poster production and adjust it as necessary.

To construct the chart, the following sequence should be followed:

1. Draw a horizontal line at the bottom of the page to represent the time axis.
2. Draw a vertical line at right angles to the zero position on the time axis on which the tasks can be listed.
3. For each of the tasks that could be started at time zero, extend a block to the time allocated for the task.
4. Where a task follows one other task, continue the block across.
5. Where a task follows two or more other tasks, draw a time block across the page from the end of the furthest of the preceding tasks. Draw dotted lines from other preceding tasks to the time block.
6. When the time block for the last task has been drawn, the minimum overall time from start to finish for poster production can be read from the scale.
7. Reassignment of the starting times of any independent tasks can be made before finalizing the chart, allocating them to periods of time when fewer simultaneous tasks are being undertaken.

The Gannt Chart will provide a clear picture of the steps involved in the production of your poster and a good grasp of the timetable. The chart should be regularly consulted and if any of the tasks are not completed as scheduled it should be adjusted accordingly.

3.2. PLANNING THE PRESENTATION

You will recall that in chapter 2 we stated that the ultimate purpose of a scientific poster is to convey scientific information and views in a visual format to an interested audience. For the message to be successfully conveyed, therefore, your poster must both attract an audience and, having done that, hold their attention for a sufficient period of time. You should select for poster presentation only those aspects of your research that are topical or controversial or that the delegates of your target meeting are likely to find of particular interest. As an astute poster presenter you must have as much concern for the audience as for the subject and your own presentation of it.

3.2.1. Targeting Your Audience

As part of your preparation, you should consider toward which professional groups your specific scientific message is aimed. You should also try to define which subgroups among these would have particular interest in your topic. These groups represent your target audience. The meeting to which your poster is submitted, the style of the presentation, the level at which the scientific information is pitched, and the eventual presentational format are all influenced by the target audience. Woolsey (1989, pp. 325–326) suggested that your poster audience can be categorized into three main groups:

> *Group* 1 comprises those colleagues, collaborators, and students who follow your work very closely. These are the people who read your publications in detail and who may even correspond with you. They will be relatively few in number and should not be considered your primary audience.
> *Group 2,* much larger, is made up of those scientists who work in the same general area as you, but on different subspecialities. This group may wish to superficially view a large number of the communications in their area. Their general knowledge of the field, however, may provide valuable suggestions and insight about your research, and this group comprises the main target audience for your poster.
> *Group 3* is made up of a large number of researchers whose work bears little or no relationship to yours. If you can attract and hold their attention for a while, however, you may be able to communicate on a more general level.

Your audience will have many other demands on their time, and a viewer will typically only spend a minute or two perusing your main points.

You are also likely to be competing with possibly hundreds of other posters, other meeting activities, and delegate fatigue. It is only those posters that have been well designed and presented in a considered professional manner that will be given appropriate attention.

3.2.2. The Venue

To help you put your presentation into context and give you a feel of the situation in which the audience is likely to find themselves, you should seek as much information as possible on the meeting venue. You should be able to glean the following information from the meeting papers.

- the mix of professional groups that the meeting has targeted
- the scope of the subject matter
- the expected audience numbers
- the total number of poster presentations
- the timetable of poster rounds
- the total time that posters will be displayed

However, you may wish to contact the meeting organizers for more detailed information on issues that may have an impact upon your poster, such as the color of the display boards, the number and size of the poster display rooms, the means and availability of material to fix the poster to the display board, and so forth.

3.2.3. Preparation for the Unexpected Opportunity to Give a "Spontaneous" Verbal Presentation

The skills required to effectively present your poster at the scheduled poster round are considered in chapter 13. However, occasionally organizers of some scientific meetings will have reserved a period of time in the proceedings for selected authors of a variety of poster displays to communicate their findings as a verbal communication. You should not shy away from this opportunity as it will provide you with greatly increased professional visibility.

The preparation to cover this eventuality need not be onerous. I would suggest that prior to attending the meeting you prepare colored slide photographs of the individual graphic elements, or alternately, colored photocopies on transparent sheets suitable for overhead projection. These will then act as a basis around which you can structure your verbal presentation. You may additionally find it useful to have photocopies of your poster text as an aide-mémoire during your talk. Even if you are not called on to present in this way at the meeting, the slides and other materials may well be useful for future verbal or written communications.

By carefully planning the production and presentation of your poster, you will be able to approach the meeting with increased confidence that your presentation will be a success. It will enable you to concentrate on enjoying the meeting and making the most of your opportunities.

4

WRITING THE POSTER ABSTRACT

The poster abstract is the part of the overall presentation that is usually destined for publication in the proceedings or abstract book of the meeting. Specific skills are required to summarize large amounts of scientific text and data into a few sentences that still adequately set the scene and convey the appropriate message. The abstract is not merely a summary of your findings. It must be able to, and indeed will, stand alone. The restriction on the number of words, the format, and the deadline for receipt will be given by the conference organizers. It is common to supply a box outline in which the abstract must be typed or printed in a camera ready format. This is the lasting part of your presentation, and you need to devote a suitable amount of time to ensuring that it maintains the same high quality as the rest of your presentation. For this reason a good quality copy should be sent for publication, avoiding faxing, as the results are often difficult to read. For casual readers this may be the only part of your presentation that is seen. You should therefore avoid the use of phrases such as "evidence will be presented," and make the abstract as representative of the whole presentation as possible.

The abstract should be tackled with the same approach as the poster. It should by its nature be concise, but nevertheless contain all the main elements. It should be lively and should be presented enthusiastically with a sense of achievement. Producing good abstracts is an acquired skill that requires practise.

Unfortunately, it is not uncommon for abstracts to be a staccato of short, sharp statements strung together. This can be avoided by careful planning of what actually needs to be contained in the abstract and by critical editing.

4.1. WHAT THE ABSTRACT SHOULD CONTAIN

The abstract must be able to stand alone. It should concisely summarize the basic content of the paper without presenting extensive experimental details. It should contain a brief statement of the need for the study, the aims, the way in which you tackled the problem, and the most important findings. Generally, the abstract should not include details of methods, unless of course the poster is introducing a new scientific method as the focus of the presentation.

Abbreviations and references should be avoided, and diagrams, illustrations, and other graphic material should not be included. If it is essential to include a reference, use the same format as given in the references section but omit the article title.

Above all, the abstract should contain enough information to clearly convey your scientific take-home message and the evidence you are presenting in support of it. As the abstract will usually be published separately in the proceedings of the meeting, it must be complete and understandable without reference to the poster text.

4.2. A SUGGESTED APPROACH

Having established what a good abstract should contain, the following is one way in which the task may be approached:

1. *Statement of introduction.* A clear statement of the need for your research should be drafted. Keep it as short as you can, but at this stage do not worry too much about the number of words.
2. *Aims of the research.* Draft a statement or statements of the aims of your study. Restrict yourself to the aims specifically related to the poster presentation and your take-home message. It is quite likely that the information you are presenting is part of a larger research program with wider overall aims. Take pains to focus only on the issues to be covered in your poster.
3. *Methods.* A statement of the way in which your aims were achieved is required. This should not detail methods but may allude to techniques. It may help to imagine that you are orally explaining to someone how you conducted the experiments. It is probable you

will use terms such as "investigated by scanning electron microscopy," or "tested by polymerase chain reaction," rather than describing step-by-step the procedures of what you did.

4. *Significant findings.* Make a list of all the significant findings you have included in your poster presentation. Again at this stage do not be overly concerned with the number of words.
5. *Take-home message.* Draft a clear, concise statement of your take-home message. Again it may help to imagine that you are orally informing a colleague of the overall relevance of your findings. This should be one sentence.

The previously mentioned statements are put together to form the initial draft of your abstract. This will almost certainly be rather jerky in style and contain far too many words. It does, however, represent the first draft, an unedited version on which you may work to produce a more polished form.

4.3. EDITING

This is the stage in which the initial draft of your abstract will be molded to produce a more accomplished piece of text that is of the size stipulated by the conference organizers. This may require ruthless revisions but you should not be deterred from this. The text will benefit enormously from judicious editing.

Condensing a sentence often requires reconstruction and rigorous editing to avoid repetition, redundancies, tautology, and the use of unnecessary words or clauses. Go through your text repeatedly, experimenting with alternate words or phrases that say the same thing but in a more concise manner. However, in your attempt at brevity be careful not to introduce terms of vague emphasis, such as *considerable, very, relatively, somewhat,* or *rather.*

Sentences may often be shortened by the judicious use of adjectives and adverbs; for example, the sentence "Every year different research findings have shown how complicated the issues in the scientific field of microbial taxonomy have become" may be written as "The science of microbial taxonomy is becoming increasingly complex."

4.4. HINTS ON STYLE

The following are some hints on style that will assist you in writing your poster abstract:

- Keep sentences short and succinct.
- Describe experiments and results in the past tense, but conclusions and generalizations drawn from them in the present.
- Separate text by new paragraphs when necessary but avoid too many as space is limited.
- Use short words in preference to long ones.
- Avoid extraneous text—keep it snappy and to the point!
- Do not include diagrams.
- Avoid jargon.
- Avoid ambiguity.
- Avoid vague expressions.

4.5. THE FINAL CHECK

When you have what you consider to be the final draft, check the text for spelling errors, incorrect data, and grammatical errors. This seems to be an obvious step, but if omitted it can lead to an embarrassing, enduring result.

Read the text aloud and listen to the rhythm created by the punctuation. Adjust it to avoid staccato lists of statements or long, breath-straining segments of text. It is also a prudent step to ask two or three colleagues to proofread the draft before submitting it to the meeting organizers.

5

WHAT TO INCLUDE

Posters generally convey scientific information by means of a mixture of printed text and visual forms of data presentation. Unlike written scientific papers, visual portrayal will predominate, often being used in place of paragraphs of text, and the poster design may also incorporate a judicious amount of color and other design features. In deciding exactly what to include, ultimately you are of course free to add to the poster anything you wish. However, your poster must be a full yet concise report of your work,and you will need to consider carefully the purpose of inclusion of any particular element. *To communicate successfully, you must focus the content of your poster on the findings that specifically support your scientific message.* You will need to include sufficient elements to ensure that your poster stands on its own, reflects your professional credibility, provides ample justification for your objectives, and gives meaning to your results and conclusions.

5.1. TEXTUAL ELEMENTS

The content of the text is an important aspect of posters, as it guides the reader through the scientific findings presented visually to the final take-home message. It links the various sections in a coherent logical fashion. However, the attention span of your audience while they are standing in a room surrounded possibly by hundreds of by other posters, as well as other distractions relating to alternate meeting activities, is predictably very short. The content emphasis should therefore be on the visual elements, the text being made as brief as possible while still maintaining meaningful intelligibility.

5.2. GRAPHIC ELEMENTS

It is likely that a variety of graphic elements could beneficially be included in your poster design, including charts, diagrams, graphs, maps, photographs, and tables. These probably represent the most important means at your disposal for visual communication of your research findings. The variety of different ways in which text and data may be presented and incorporated into the poster are considered in detail in chapters 6 and 7, respectively.

5.3. ADDITIONAL INFORMATION

Supplementary information, including references, contact details, and so forth, may also be made available and occasionally integrated in the overall poster design. This aspect is not considered here as it is covered in chapter 8.

5.4. THREE-DIMENSIONAL MATERIAL

Display materials, such as preserved specimens, cultures, and others, may be attached to the poster. This may be accomplished by means of double-sided adhesive tape, superglue, or by the use of Velcro strips. This may provide a novel attraction to your poster but caution is advised. The material must have relevance to the overall theme and its inclusion must represent the most effective means of conveying particular information. It must not appear to be merely stuck on as an afterthought, but must serve a purpose and be intimately integrated into the overall poster design.

5.4.1. Large or Heavy Appendages

Where it is not possible to attach an appendage to the poster, for instance in the case of a particular piece of apparatus or a large, preserved specimen, it may be possible to display this separately from the main poster, on a small table if available or on a chair. However, it would be prudent to discuss such needs with the conference organizers, who will almost certainly require a convincing justification of the need. Chairs and tables are unlikely to be readily allowed in the body of a room assigned to poster presentations as they will block aisles. However, it is not uncommon to have tables and chairs around the walls of the room. If contacted early enough, the organizers may be persuaded to arrange for your poster board to be positioned conveniently next to these. Never position anything directly in front of your poster, however, as this will only serve as a distraction and may actually conceal part of the poster. Indeed, there is a danger that the additional article may steal the show almost totally from the poster. Unless such materials are vital to the presentation or specifically what the poster is about they should not be included.

5.4.2. Live Specimens

It is almost invariably inappropriate to display live specimens, although there may be occasion to display live plants. This is fraught with difficulties, however, particularly in maintaining the viability of live specimens in transport. Again such displays are to be discouraged unless absolutely necessary. Additionally, there may well be restrictions on movement and importation of certain live materials. Those attempting such a procedure should check with the appropriate authorities in their country of origin and the country of destination as to current legislation.

5.5. DESIGN ADJUTANTS

Some design aids are commonly used with good effect in poster designs. Illustrations relating to the subject of the poster, but not directly to any particular element, when superimposed on the backing material may effectively contribute to the overall design. Arrows in particular are useful in directing the information flow where the reading route is otherwise not easily discernable. However, you should use such adjutants sparingly, taking care not to disrupt the flow of open space or distract from your main textual and graphic elements.

5.5.1. Humorous Elements

The use of humorous textual or graphic elements is sporadically seen in scientific posters. It is usually included in an attempt to break down barriers in making contact with the viewer, or to lighten a particularly enigmatic section. Its use, however, is fraught with difficulties. At best you may amuse a portion of your viewers who share your sense of humor, and at worst you may offend those who do not. Your attempt at humor may be interpreted as buffoonery and as an annoying distraction that brings into question the credibility of your work and your standing as a seriously committed professional. I would therefore suggest that the possible gains of incorporating humor into your scientific poster do not warrant the risk of misinterpretation, and I would strongly recommend that you resist the temptation to include such material.

6

PRESENTATION OF TEXT

The textual elements of your poster will inherently have an important role in communicating your scientific findings, but they will also have a considerable impact on the overall design. Text must be presented in a professional manner, either typed or printed, but textual elements should never be handwritten. In this chapter we consider the ways in which text can be manipulated to have a more pleasing appearance that tempts the viewers to read, how it can be made to fit into specific areas that are available in the poster, and how it can be effectively incorporated as part of the overall poster design.

6.1. STRUCTURE OF THE TEXT

When writing the various sections of text, I would suggest that you may find that it is helpful to consider it as "expanding the abstract rather than condensing the paper." Large blocks of text should not be used. Paragraphs should be kept short, preferably no longer than 10 to 20 unbroken lines. The organization of the text into manageable and digestible components of the overall poster is considered in detail in chapter 9.

Whatever order you choose for the arrangement of the material, it is important that you maintain presentational unity. Blocks of text when viewed from a distance appear as blocks of gray shading. The density of the text will influence the depth of gray. The margins, line spaces, typeface, and other elements will affect the "openness" in the segment. The line length and number of lines will demarcate the size of the area, but the overall shape will be influenced by the chosen justification and by whether any of the text has been wrapped around a graphic element. You must decide on the typeface, size and weight of type, hierarchy of headings, and other elements and use them consistently throughout.

6.1.1. Choice of Typeface

The number of typefaces available today is staggering and making a choice can be confusing. The limitations in the first instance may be the software of the computer or the word processor, or the faces supplied with the typewriter, to which you have access. However, it is likely that at least one of these will be a serifed face and one other a sans serif. It is important, however, that you choose one that your poster viewers will find easy to read. Legibility is an obvious starting point. Laser printer fonts based on classic type designs are most legible and, depending on the font chosen, are generally more compact than typing. Text must be enlarged sufficiently to be easily readable at the distances likely to be encountered at a poster session (see Table 6.1.). Legibility, however, is a product not only of the basic shape and size of the letters, but also of the measure (width of the setting) and the leading (space) between the lines. The most obvious difference between one typeface and another is the presence or absence of serifs (the crosslines at the ends of the main strokes in each letter).

SERIF SANS SERIF

Generally, serif typefaces, such as Times, are easier to read than sans serif faces, such as Helvetica. The crosslines of serif help to guide the reader's eye. I would therefore recommend the use of these for the body of the poster text. Open, round letters are preferred to condensed, small-bodied ones and typefaces such as Courier, Prestige, and Times are suitable choices. Sans serif typefaces may be used for shorter labels and textual elements such as main headings, but you should avoid inappropriate ornate or novelty fonts such as **Old English**.

6.1.2. Size and Weight of Type

Most typefaces are available in a wide range of "weights," or thickness of line; for example:

normal, **bold,** *italic.*

Italic is generally more suitable than bold for emphasis within sentences, because it makes a distinction without being too obtrusive. Bold can be used effectively for a few words at the start of key paragraphs to act as a minor subheading. Appropriate text characteristics for use in the various components of your poster design are shown in Table 6.1.

Variations to the basic typeface, for example bold or italic, are extremely useful devices for design emphasis of particular words or phrases. It is likely that you will also use bold or italic styles for headings or captions. Within the text, small capitals for abbreviations may be less obtrusive than capitals. Acronyms (Nato, Unesco, etc.) can be set in upper and lower case if preferred.

6.1.3. Text Alignment

You will have to decide on the alignment of the text. It is possible to align text in several ways; justified, flush-left/ragged-right, ragged-left/ragged-right, or centered. This is principally a matter of taste, although it is my experience that type in lines of uneven length (flush-left, ragged right) scans better than justified (even line length) type when read from a distance. Whatever alignment style you select for the main body of your text, it must be used throughout the poster to maintain a unified appearance.

6.1.3a. Justified Text. If lines are justified, the space between words is varied to drive lines out to a common length, as in the main text of this book. This can sometimes lead to excessive space, particularly if lines are short, and if these spaces link up between successive lines they can appear as rather disagreeable "rivers" of white flowing down the text. Justified text is often used in the body text of posters and may project a serious or classic image.

Table 6.1
Appropriate Text Characteristics for Components of Posters

Textual Element	Reading Distance (m)	Type Size (mm)	Type weight
Main Title	3–4	30–45	Bold
Authors and affiliations	3–4	25–30	Bold
Main headings	2–3	10 (36 pt)	Bold
Subheadings	1–2	5 (24 pt)	Normal or bold
Main text	1–2	5 (24 pt)	Normal

6.1.3b. Flush-left/Ragged-right. Flush-left/ragged-right is characterised by equal word spacing and lines of unequal length. This form of alignment often creates "open" pages because each line of type contains a slightly different amount of white space at the right border. This type of alignment is generally considered to be more informal, friendlier, and more open than justified. This paragraph would appear thus:

Flush-left/ragged-right is characterized by equal word spacing and lines of unequal length. This form of alignment often creates "open" pages because each line of type contains a slightly different amount of white space at the right border. This type of alignment is generally considered to be more informal, friendlier, and more open than justified.

6.1.3c. Ragged-left/Flush-right. Ragged-left/flush-right produces the opposite effect to flush-left/ragged-right. This alignment, however, does not make for easy reading and is therefore usually restricted to particular design effects. This paragraph would appear thus:

Ragged-left/flush-right produces the opposite effect to flush-left/ ragged-right. This alignment, however, does not make for easy reading and is therefore usually restricted to particular design effects.

6.1.3d. Centered. Text copy may be centered about a vertical midline and to a maximum width. This style may be useful for presenting lists that do not have bullet points and for titles and subtitles. If centered, this paragraph would appear as shown below:

Text copy may be centered about a vertical
midline and to a maximum width. This style
may be useful for presenting lists
that do not have bullet points,
and for titles and subtitles.

6.1.4. Headings and Subheadings

Headings and subheadings should be included in the poster to help direct the viewer and to organize the communication. However, you should avoid using too many as this may simply serve to confuse the reader. A hierarchy of styles should be employed; for example, by making use of different type sizes and weights, thus implying the same degree of emphasis each time they are used.

Heading, subheading, sub-subheading

It is common to use the same typeface for headings and subheadings as it produces a professional looking result that maintains the unity of the poster. However, for particular emphasis a different typeface to the main body text may be used, for example, a sans serif heading over serif body text, thus:

> **Sans serif heading**
> Serif body text,serif body text,serif body text.Serif body text,serif body text,serif body text.

Variation and further emphasis may be obtained by adding rules above and/or below, or by containing the heading in a box. Color options are also a possibility and are considered in chapter 8. Use of centered justification may provide yet a further option.

6.1.5. Line Length and Spacing

The human eye tends to tire after it has ranged over too many characters in one line, so you should ensure that lines are not too long for comfort. For posters, lines should be single spaced and should contain between 20 to 60 characters. You may find it necessary to resort to a double-column setting to reduce the line length to manageable visual proportions. Excess line spacing also contributes to difficulty in reading. As a rule of thumb, line spacing should be approximately 20% of the type height. A line space can be placed between paragraphs, or the first word can be indented.

6.1.6. Notes and References

Superior figures may be inserted at the ends of words, phrases, or sentences to which notes apply, thus[1]. Alternately, references can be denoted by numbers in brackets within the text. The latter may, however, interrupt the flow of the text if you intend to add a considerable number of notes or references. If it is important that notes are referred to easily, they should appear in smaller size type as footnotes to the specific section. If the notes are to provide backup information they should be consigned to a separate notes section (see chapter 6).

Cross-references. These may refer to illustrations, other text passages, tables, or notes. They may contain the essential information in parentheses, or may include the words *see, see also,* or cf. (compare). The more specific the cross-reference, the better. Avoid *see earlier, see above,* and *see later,* which are too imprecise. However, avoid overuse of cross-references as it is likely to ruin the continuity of information flow.

6.2. VARIATIONS IN STYLE

You are of course at liberty to vary the style of the text used to add panache and individuality to your poster. Openings at the start of new sections or paragraphs, for example, are sometimes marked by a dropped or raised capital, and/or occasionally by setting the first few words or line in capitals, as shown in Figure 6.1.

You may wish to put particular emphasis on some pieces of text and this can be achieved by a variety of means, including separating the words from the main text; setting in a bold font; increasing the size of the type; the use of bullets, asterisks, or bold subheadings; underlining; setting in a box; and using tints enclosed in boxes. The use of color to emphasize segments of your poster, including textual elements, is considered in chapter 9.

6.2.1. Lists

If sections of text cannot be reduced to short paragraphs, these should at least be broken up with a new heading, picture, table, figure, or other illustration. Concise lists are often a convenient alternative to continuous

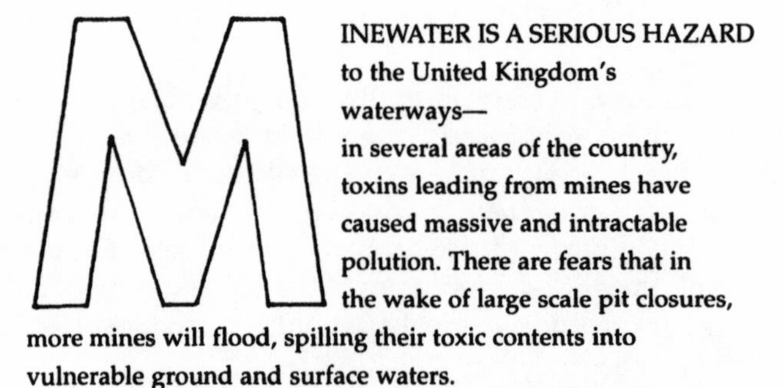

Figure 6.1. An example of a stylistic opening at the start of a new section. The design makes use of an enlarged dropped initial capital, and continuation of the use of capitals for the first line, to add interest to the text.

runs of text, particularly in the aims and conclusion sections. *Use lists when possible to reduce continuous runs of text.*

The content of lists needs to be clear and concise. Each statement should consist of a single matter of fact or area of concern. A list is best presented as a separate element removed from the main body of text. Emphasis can be added by using symbols such as bullet points or stars; for example, •, *, o. If the separate points have a logical sequence, these may be appropriately numbered or alphabetically labeled. Lists need not be boxed but can instead be indented to the main text to again signify a separate element. Lists are best when they are concise, presented as separate elements, and indented to the main text.

6.3. FINE TUNING AND MAKING IT FIT

Once you have chosen your typeface and size, you may wish to calculate the approximate number of lines of text you will be able to fit in a particular area of your poster to guide you in drafting. This can be easily calculated once you know that there are 72 points to one inch. For example, if, allowing for top and bottom margins, the text area is, say, 8.5 inches deep, and the text is to be set in 24 pt with interlinear spacing of 4 pt, each line and the space following will occupy 28 pt. In 8.5 inches there will be 612 points. This means that 22 lines of text of this size can be accommodated in that area of the poster (612/28).

If it is necessary to save some space in a particular area you should not do so by reducing the size of the text itself, as this will upset the unity of the poster. It may be possible, however, to save space to enable the different elements to fit better by making a few of the illustrations smaller. Alternately, perhaps the bibliography could be set in a double column to save space. With increased experience such maneuvering can be made with confidence.

A certain amount of space can be conserved by fine tuning the text; for example, by reducing the line spacing between large sized text, or by "kerning." The technique of kerning involves deliberately reducing the spacing between specified characters, leaving the rest of the settings the same.

6.3.1. Kerning

This technique is frequently used with certain letters, such as "Yo," "Te," and "LY." When these are printed, there is often too much space between them, compared to the rest of the letters. Kerning corrects this problem. If properly used, it can improve letter fit and legibility, and it is particularly useful with large type sizes such as those used for the poster title or for main headings. The following example shows the word "partly," a) printed normally; and b) printed after kerning.

a) PARTLY b) PARTLY

6.4. PROOFREADING AND CORRECTIONS

It is essential that your final text is read and checked, ideally by at least three pairs of eyes. You should ensure that all matters of fact and of consistency have been resolved and that there are no spelling or other typographical errors. The checking of text needs to be methodical, line-by-line, and word-by-word. Even if your computer or word processor contains a spelling dictionary capable of checking a range of spellings, this will only reduce the room for error. It will not eliminate it altogether. It is difficult to spot such errors on a visual display unit and a hard copy printout is recommended for proofreading.

This stage of production of your textual elements is not the time to introduce "second thoughts," as even relatively minor textual changes may alter the overall appearance and fit of the elements in your poster display.

We have now considered some of the major factors influencing the textual elements of your poster. You should be now be able to confidently produce text that will be legible, informative, concise, fits the available area, and that effectively contributes to the overall design of your poster.

7

PRESENTATION OF DATA

Scientific posters rely heavily on visual presentation of research data and trends for effective communication. Although certain simple statistical facts and comparisons can be embodied within a short narrative text, for example, "About 3 million blood components are issued by the UK transfusion services each year. Only eight cases of infection transmitted by transfusion were reported." A point is often reached when, for ease of understanding, such information is best siphoned out of the text and presented separately. A more readily comprehensible and visual format for the data is required. This may be achieved by organizing and analyzing your data into a form that can be represented by graphic techniques.

For those of you with access to a computer, programs are available that not only enable graphs, bar charts, and pie charts to be constructed, but are also able to convert information held in tabular form into graphs and charts and vice versa. Alternately, however, graphics can be prepared manually as line art work, and the lack of access to computers should not be seen as a barrier to producing highly effective graphic elements. To illustrate this point all of the graphics in this chapter were originally prepared manually. Indeed, the use of computer-generated graphics can have some disadvantages, as they encourage the use of overly complicated designs. *To be effective in posters, graphic elements need to be kept simple.* They should enable the viewer to rapidly translate the element into a clear scientific message.

Statistical information can be presented using a variety of graphic techniques, the principal as being diagrams, tables, graphs, photographs, maps, and charts. We will consider these techniques in detail but should first give some consideration to the nature of "raw data" and the stylistic aspects of graphic elements.

7.1. RAW DATA

After having completed an experiment or set of experiments, it is likely that at least two sets of figures representing the scores or outcome of experimental and control participants on particular tests will be obtained. In this form, the data is considered to be raw data as nothing has been done to extract any meaning from the numbers. Unless the number of scores in each set is small, it will be very difficult to get the impression of what the data mean merely by inspecting a long column of figures. It will require considerable effort to extrapolate even the simplest pieces of overall analysis, such as the highest score and the lowest score. It will additionally be almost impossible to analyze the spread by looking at the haphazard arrangement of scores. It is therefore most unlikely that your poster will incorporate raw data in its design.

7.2. STYLISTIC ASPECTS OF GRAPHIC ELEMENTS

The presentation of data, whether it is in tabular or in graphic form, will have a considerable impact on the overall appearance of the poster. Aspects of the design and format of your poster are considered in detail in chapter 8, but whatever your design it is likely that you will have a center focus, and that is most likely to be a graphic element. You will, however, need to be selective with the number of charts, graphs, and so forth, as too many will be overpowering stylistically and with merely serve to confuse your viewers. Additionally, many of the points identified as being of importance in textual elements in chapter 5 also relate to numerals and to letters included in graphic elements. The choice and mix of graphic elements must both effectively convey your message and be exciting visually, but their size and style must also conform to the overall design of the poster. A unified style of presentation must be maintained.

The number of illustrations should be restricted to no more than eight to ten per poster. Final figure and table size will depend on the complexity of the information but they must be of an appropriate size to maintain legibility.

Graphic elements need to be designed and manipulated to add verve to your poster. Effective tables and charts breathe new life into scientific data. They make it easy for your poster viewers to instantly visualize and assimilate the message you wish the data to communicate. *Graphic elements should not be buried within columns of text.* You should surround graphic elements with whitespace so they clearly stand out.

7.2.1. Positioning and Organization of Graphic Elements

The precise position and dimensions of your graphic elements will largely depend on your overall poster design and format (see chapter 8) and the relative importance of the element in communicating your scientific message. However, some stylistic principles will be relevant whatever your choice of final design.

With the exception of a center piece, graphic elements will generally follow the same columns as your textual elements and will be aligned with each other and with the margins of adjacent text columns. Always ensure that you provide plenty of clear space between a graphic element and adjacent text. Even a small element surrounded by whitespace will communicate its intent better than a larger one surrounded by text.

Graphic elements should always be referred to in the text and whenever possible should be positioned directly following or opposite the position where they are referenced. When this is not possible, effective numbering, labeling, or caption heading should enable them to be readily located, and they should at least be in close proximity to the text referring to them.

7.2.2. Numbering

Your figures and tables should be numbered for ease of location. Reference can then be made to them in the text by use of "See Table 1, 2, 3, etc. or Figure 1, 2, 3, etc." The value of this becomes particularly apparent when the element is not positioned directly following its textual reference or when cross-references are being made.

7.2.3. Headings

Each graphic element should have a main descriptive heading, and in the interests of maintaining poster unity, these should all follow the same style in upper and lower case, either normal or bold, keeping initial capitals to a minimum. The heading should be as short as possible so the reader can see at a glance what the table contains.

7.2.4. Captions

Captions should either appear beneath or alongside individual illustrations. However, on occasions where there are several illustrations of the same subject appearing in the same segment, it might be sensible to consolidate the captions into one to avoid unnecessary repetition. If illustrations and captions are numbered, the consolidation of captions presents no problem, because the numbers for the captions will relate to the numbers for the illustrations. If the illustrations are not numbered, then the captions must be preceded by *above, below, center, right, left,* and so forth unless it is immediately obvious which is which.

Captions should be clearly distinguishable from the main text. To achieve this they can be set in a smaller size. You should avoid repeating in captions information that appears alongside in the main text. It may be preferable, however, to complement the information in the main text by giving additional information in the caption.

7.2.5. Legends and Labels

Captions appear outside of illustrations, the terms "legends" and "labels" are used here to refer to explanatory text that forms part of the graphic element. The type size relates to the size of the diagram in question. Text should neither be too small so it is overwhelmed by the illustration nor too large so it dwarfs the illustration. When parts of a diagram are being labeled, it must be clear as to which text refers to which part of the illustration. You should try to ensure that legends, where possible, are set horizontally. If legends have to be set vertically or at an angle, make sure that they do not involve the poster viewer twisting his or her head from side to side to read them.

7.2.6. Three-Dimensional Effects

Three-dimensional effects can be used to add depth to most graphic elements. Minor modifications in the appearance of a chart can often pay big dividends in the communicating power of your chart and can also make improvements in the appearance of your publication. However, care should be taken to avoid misinterpretation of the scientific message by ill-advised distortion of graphic elements. Three-dimensional pie charts, for example, may place a greater emphasis on lower segments than is appropriate.

7.3. CHOOSING WHICH GRAPHIC ELEMENTS TO INCORPORATE

It is best to decide at the outset which pieces of information are of greatest importance in getting your message across. You will also need to decide which of the graphic techniques available would be the most effective in each case, ensuring that they remain conducive to your overall poster design.

Generally, simple facts and comparisons of quantity or trends over time are the preferred type of information for depiction in a poster graphic element. Complex data, which is best presented using conventional tables, should not be included in your poster design but may be provided if necessary in a separate information sheet (see chapter 10).

Several basic types of charts exist, as well as many variations. The particular graphic style you use is of course your personal choice, but undoubtedly each style will portray the same data in a slightly different way, often with a different focus or emphasis. To assist you in making your decision as to which graphic technique may be most appropriate for any particular piece of information, the rest of this chapter is taken up with considering in more detail the principal options and with providing you with a compendium of examples of each style.

7.4. BASIC GRAPHIC ELEMENT DESIGNS

The most popular basic graphic elements that have proved to have the best communicative power in a poster display include tables, graphs, and a variety of different charts.

7.4.1. Tables

Tables with a limited number of data points can effectively convey a scientific message. However, if more than 20 items are to be contained within the table, it would be preferable to consider alternate forms of graphic illustration.

Tables consist of rows and columns made up of figures, words, or a mixture of both that are separated and linked by rules or spaces. The columns have descriptive headings, and the rows sometimes also have descriptions on the left-hand side. You will need to consider the best alignment for the table with respect to the overall poster design and to the way in which the data needs to be assimilated by the viewer. Vertical rules or spaces tend to stop the eye from running horizontally. Conversely, horizontal rules or spaces tend to stop the eye from running vertically. The shape of the table can usually be adjusted if necessary by rearranging the data, so, for example, a wide table can be turned into a narrow one by switching columns and rows. Emphasis can be given to a column or a row by highlighting it with color, but overuse may nullify the value.

7.4.1a. Numbers. Many tables contain numbers. If they are large numbers, all of the same order (thousands or millions), the 000s and 000000s can be omitted, provided that the table states that figures are given in thousands or millions. If there is a mixture of large and small numbers, this is more difficult because small numbers would then have to be shown as decimals of the unit of measurement. If all of the items in a table are in the same units of money, weight, or other quantity, the tables can state this once. The more simplified the data appears in the table the easier the scientific message will be assimilated.

7.4.1b. Rules. Rules are often used to divide tables horizontally, but columns tend to be divided by space rather than by vertical lines. Thick rules, thin rules, and rules of different colors can be used for different emphasis. Horizontal rules are useful to mark the start and end of a table to distinguish it from the main text, and within the table itself rules can be used to organize and structure the material.

7.4.1c. Dittos and Gaps. Ditto marks should be avoided, and the relevant term or number should be repeated. Gaps can be left if there is no information available, though it may be preferable to insert "N/A" to indicated that the information is not available or not applicable.

7.4.1d. Notes. Notes to tables should be kept to an absolute minimum. When unavoidable they can be set at the foot of the table and can be keyed against the relevant items by making use of either symbols or superior letters (superscript).

Column headings and side headings should be short. Ideally, there should be noticeable typographic contrast between the information used to identify the content of rows and of columns compared to the information contained in the rows and the columns.

Spacing is critical to the clear illustration of data in tables, and text that is crammed next to adjacent grid lines or text that extends too close to text in the next column should be avoided.

7.4.2. Graphs

Graphs effectively display trends, but to be effective they need to be kept simple. The scale should be such that the crucial movement is accentuated, as is illustrated in Figure 7.1. If several sets of figures are to be plotted, it will be necessary to use different graph lines. These should be clearly differentiated in bold, solid, broken, or dotted lines or in lines of different colors if the distinctions are to remain clear (see Figure 7.2). The number of lines in a graph should be limited to no more than three whenever possible.

As in other graphic and textual elements, it is important to maintain

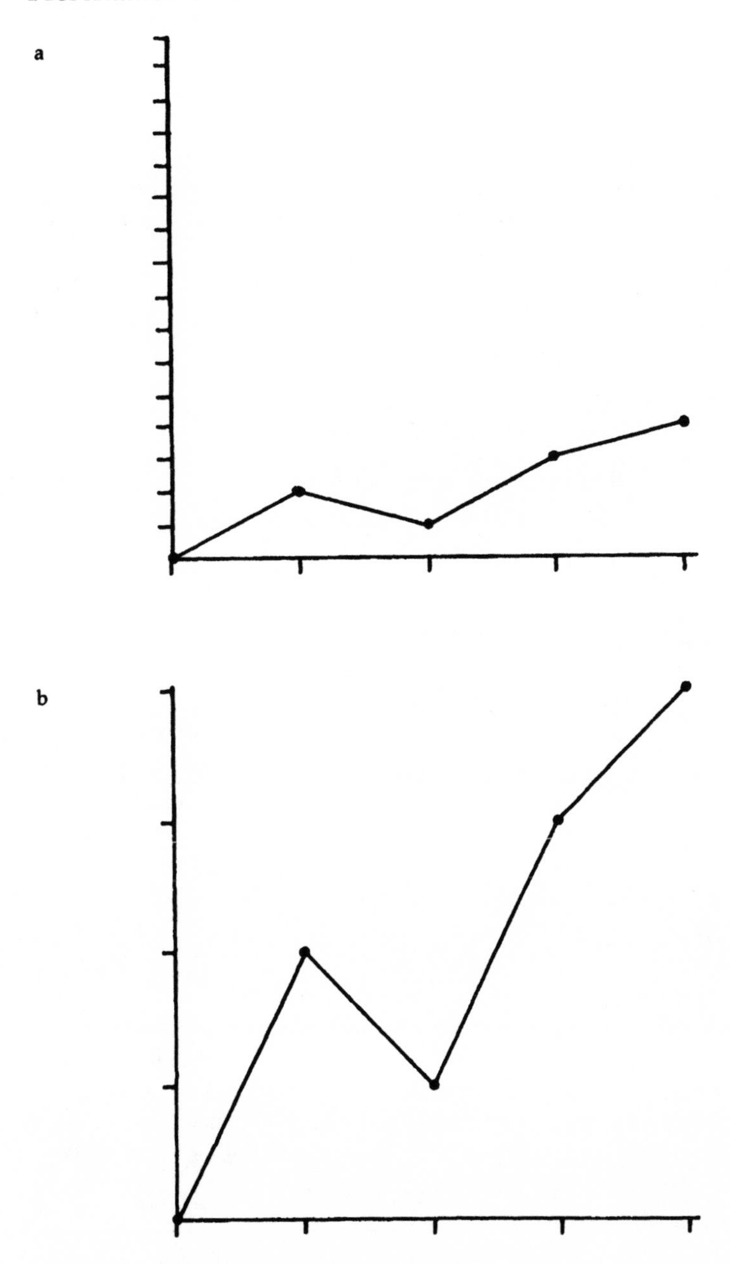

Figure 7.1. Graphs showing the value of stressing the information to be communicated. Graphs a) and b) both display the same information. However, in b) the scale on the y-axis has been adjusted to accentuate the critical movement to more clearly communicate the trend.

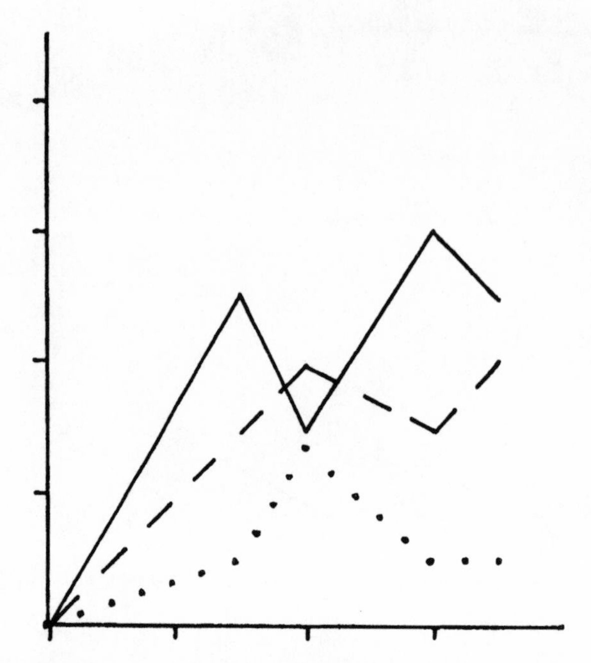

Figure 7.2. Different graph lines should be used to maintain clear distinctions. To maintain clarity no more than three lines should be shown on a graph for poster display.

consistency, including the labeling from one graph to another. If a data line is represented by a color or a shape in one graph, the same demarcation should be used in subsequent graphs featuring that data on the same poster.

7.4.3. Charts

Information shown in graphs can also be presented in the form of charts. These allow words and figures to be incorporated into the charts themselves, which can help to enliven the poster and put points in a snappy manner. There are a variety of different styles of charts that may be adopted, but by far the most popular are bar charts and pie charts, each with a number of variations. Each type of chart displays data in a different way, but to convey your message effectively, whichever type is used must be kept as simple and uncluttered as possible.

7.4.3a. Bar Charts. Bar charts work well on poster displays and are subsequently frequently incorporated into poster designs. A variety of different modifications of the basic bar chart may be adopted, as illustrated by

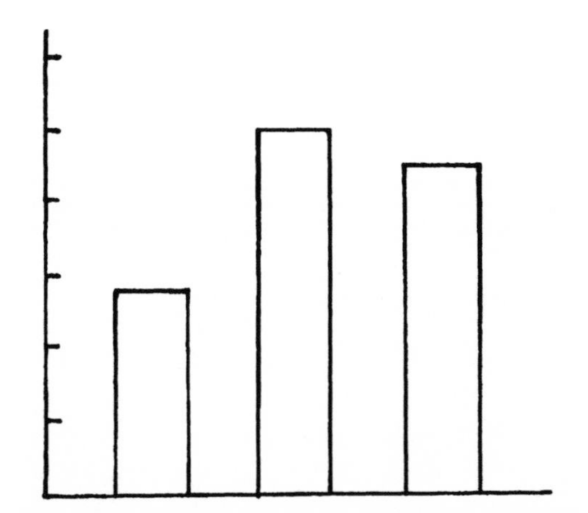

Figure 7.3. Basic bar chart. Simple bar charts make it easy for readers to compare data that changes over time.

Figures 7.3 through 7.6. These enable selection of the most advantageous features in emphasising the desired interpretation of information.

The number of bars illustrated are best restricted to no more than six, which can be readily differentiated by shading and hatching. However, complicated cross-hatching should be avoided. Color differentiation of bars can be used with good effect.

Three-dimensional versions of bar charts, and those presented at an angle, can also be most effective, as shown in Figures 7.7, and 7.8.

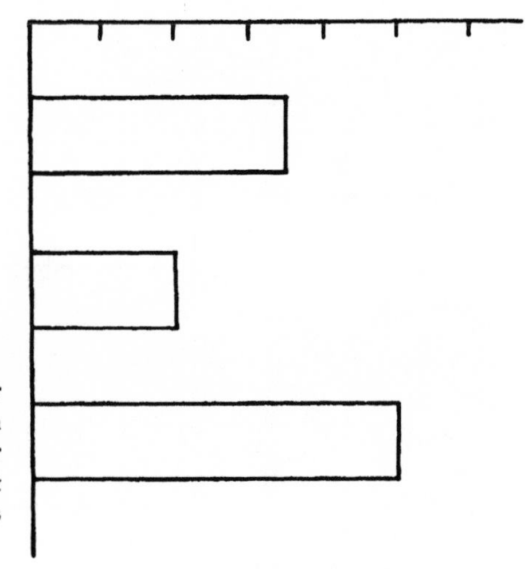

Figure 7.4. Variation of a basic bar chart. The information is presented in a horizontal format. Horizontal bar charts display comparisons that do not involve time (or when the x-axis labels are too long).

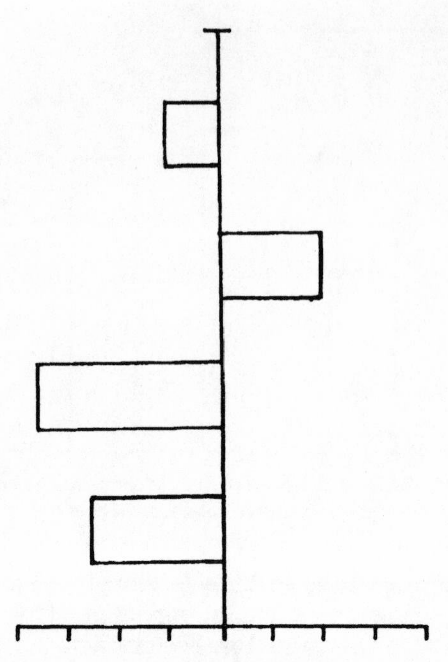

Figure 7.5. Variation of a horizontal bar chart. This chart places the y-axis midway along the x-axis to enable negative as well as positive readings to be illustrated.

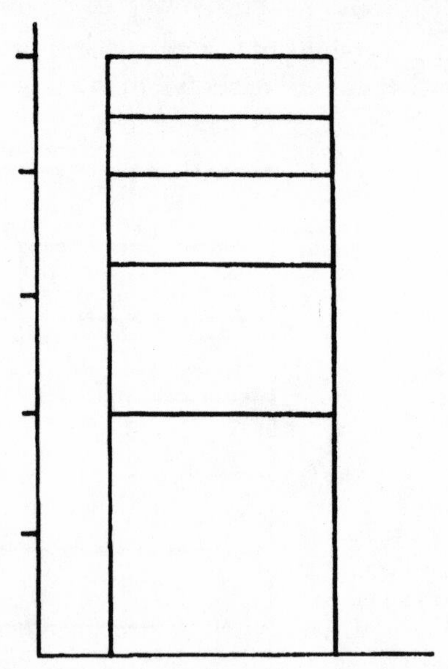

Figure 7.6. A 100% stacked vertical bar chart. These charts enable illustration of the relative contributions that make up the total.

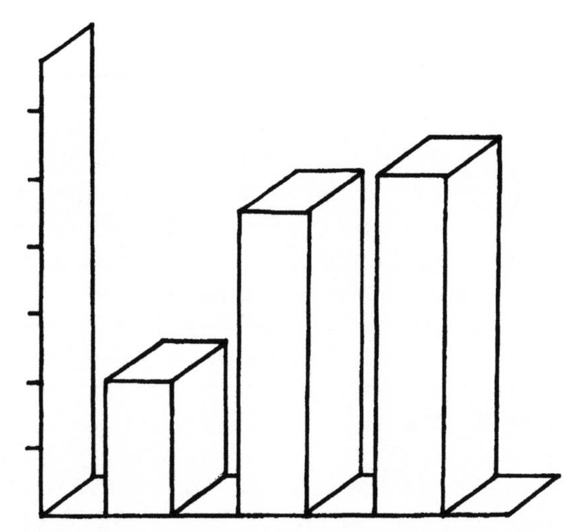

Figure 7.7. A three-dimensional vertical bar chart. These charts add interest to the information being illustrated, and can be integral to the main focus of your poster.

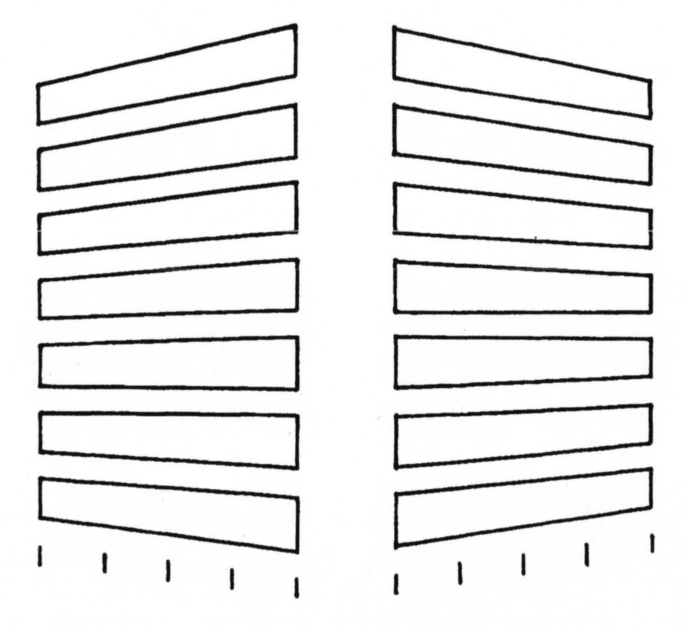

Figure 7.8. An angled version of a horizontal bar chart. Placing a chart at an angle can have a very pleasing effect. This chart shows the effect of have two sets of horizontal bars placed at an angle. The design gives the impression that the center portion is projected forward, and adds emphasis to any information on the y-axis.

7.4.3b. Pie Charts. Pie charts are also a popular means of illustrating data and are very effective in displaying how parts relate to a whole. They depend for their effect on there not being too many subdivisions, and I would not recommend use of more than six. The most important sector should be placed at the 12 o'clock position and if all the components are of equal value, arrange them from the smallest to the largest or vice versa. The segments can be shaded or colored if required to help in distinguishing them or to signify their importance, but with clear labeling this may not be necessary. You should remember to always place your words horizontally, not so they follow the shape of the pie.

Pie charts can be modified in many ways, and Figures 7.9 through 7.14 show examples of some of the variations. These variations include basic pie charts, three-dimensional charts, shadow charts, and charts in which slices or segments are pulled away from the chart. The last technique, shown in Figure 7.13, is often used to draw attention to an important slice and has a pleasing stylistic effect. Occasionally the pie chart is linked to other charts or tables to display the data making up a particular slice, as shown in Figure 7.14.

7.4.3c. Other Charts. There are a variety of charts that are occasionally used to good effect on posters; for example, Figure 7.15 shows the use of

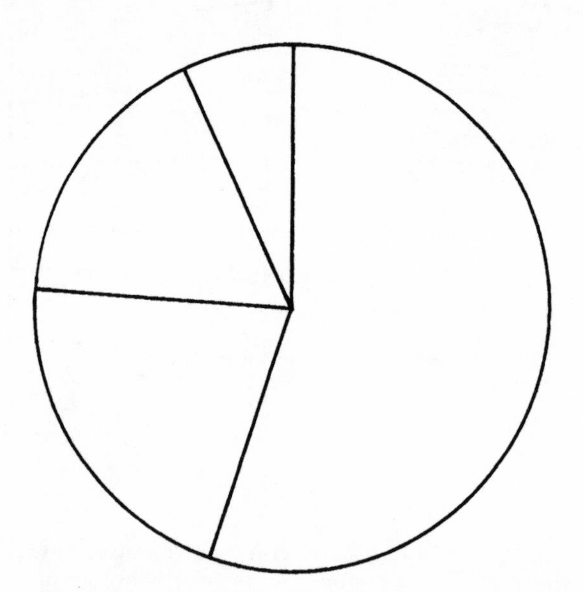

Figure 7.9. Basic pie chart. Pie charts offer a stylish way to illustrate the relative contributions that make up the total.

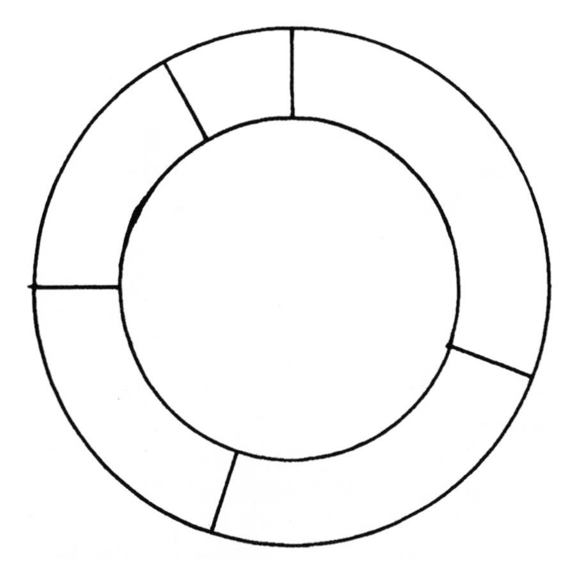

Figure 7.10. Variation on a basic pie chart. This variation on a basic pie chart gives a more "open" feel to the illustration, which may contribute to your overall design.

overlapping ovals to illustrate comparative data. Others include scatter charts, which display the distribution of data through time; maps, which may be particularly useful in illustrating epidemiological data; and radar charts, which display the distribution of data. Each may have a place on the right poster, and only you can ultimately decide the best way of illustrating your own particular scientific data.

7.5. PHOTOGRAPHS

Photographs will help to make the poster attractive, but must be directly relevant to the subject of your poster and your scientific message. They

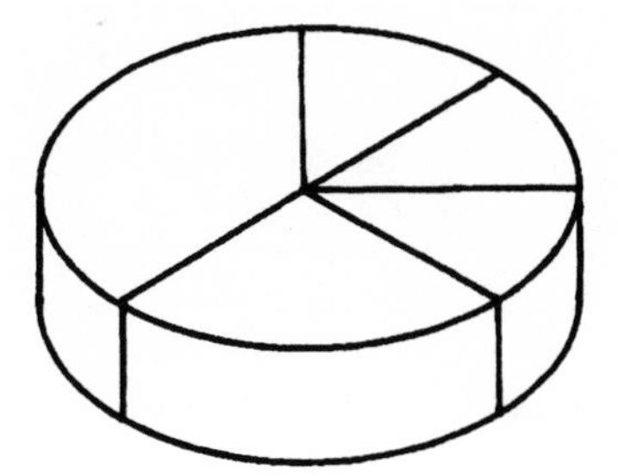

Figure 7.11. Three-dimensional pie chart. The segments of three-dimensional pie charts may be shaded, colored, or left uncolored but they should be labeled. This figure shows the emphasis that is placed on the lower segment.

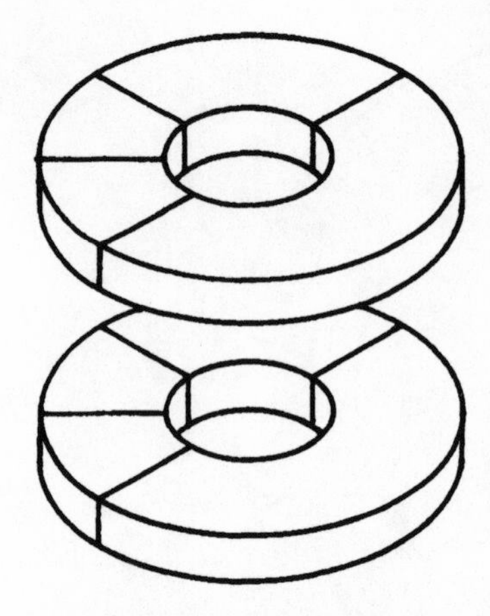

Figure 7.12. Shadow pie chart. Further information may be added within a single figure by adding identical "shadows." The segments directly below each other may be labeled with the additional information.

must be clear and large enough to be easily discernable when viewed at between 1 to 2m. A large color, matte-finished photograph, snappily illustrating an important aspect of your research, can be a great asset in attracting attention to your display. However, a large number of small glossy ones can be most distracting for the viewers. It is better to enlarge the area of a photograph that contains some particular feature so it can be easily seen than to leave the picture with a lot of extraneous material in view.

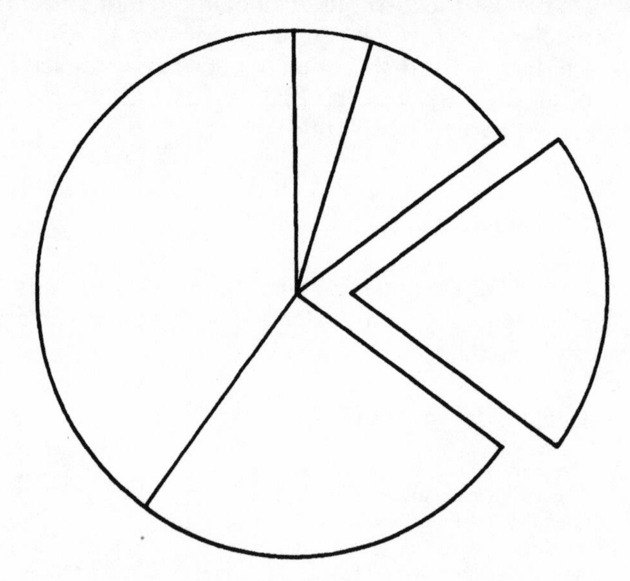

Figure 7.13. Pie chart with slice pulled away. Pulling a slice from the pie adds emphasis to that particular segment.

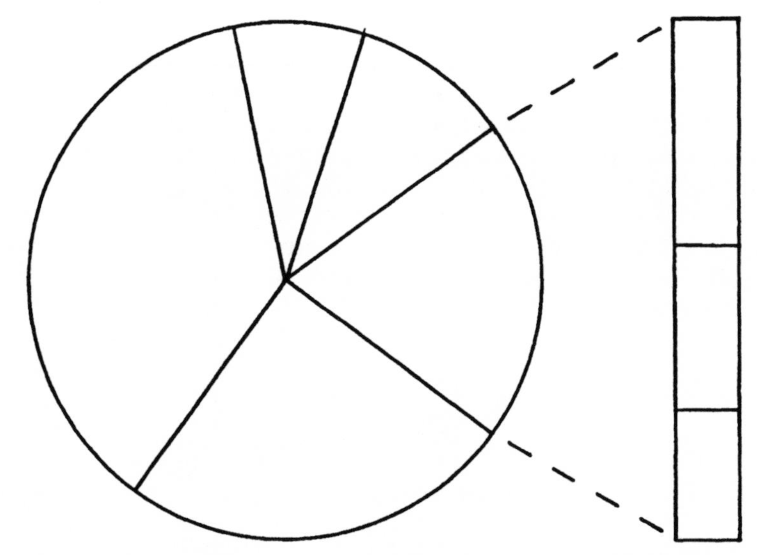

Figure 7.14. Pie chart linked to a bar chart. Where you have too much information to add to a segment label, one technique, as shown in this figure, is to add the information on a bar chart that is linked to the particular slice.

If color photographs are not available or not appropriate, good quality black-and-white ones may equally suffice. If necessary, these may be colored with transparent markers and adhesive overlays to call attention to certain details.

7.6. MAKING IT FIT

There is more latitude to adjust the size of graphic elements to make them fit the available space on the poster than with textual elements. However,

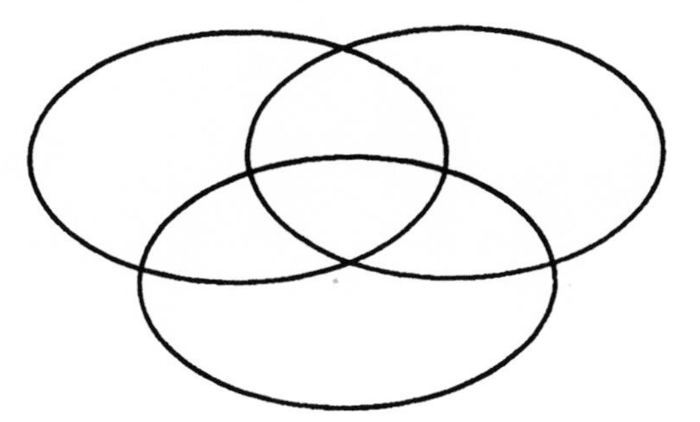

Figure 7.15. Overlapping ovals to illustrate comparative data. Overlapping ovals or circles is a simple way of illustrating comparative data. the segments may be shaded, colored, or left uncolored but they should be labeled.

you should remember that by altering the size you may alter the impact a particular element may have. You must also bear in mind that by adjusting the size of a chart or graph, you may also need to readjust the size of the text associated with the particular element. If you reduce the size of the graphic element you may need to increase the text and the numbers contained within it to maintain legibility. Conversely, if you enlarge the graphic image you may need to reduce the associated text and numbers to maintain a unified and balanced presentation. You should always ensure that the legend and any data labels used in charts are as easy to read as adjacent text.

8

THE FORMAT AND DESIGN

The best posters usually make one to three main points. These constitute your take-home messages with which you hope your audience will depart after viewing your poster. Simplicity and clarity are vital. The format and design of your poster should be built on the basis of conveying your take-home messages in the most effectual manner.

This chapter considers the poster as a whole in terms of the format and design. It reviews the options open to you and offers guidance on incorporating the material available into the overall design.

8.1. ORGANIZATION WITHIN THE POSTER

Posters are traditionally split into sections following the format of written scientific papers; title,summary, introduction, methods, results, conclusions, discussion, and references. This is the general organization within the poster that your audience expects to see. This does not of course mean that you must stick rigidly to this formula, but the order does have a logical flow and the unexpected, if too dramatic, may confuse the reader. If an unexpected format takes too long to decipher it is likely not to be read.

8.1.1. Title

The title is the first thing that most people viewing the posters will seek and read. *The title announces you and your work.* It is important that, even from a distance of 6 to 10m, it is legible, assertive, clear, and eye-catching. The poster title should be worded in exactly the same way as it appears in the abstract published in the meeting's papers. The names of the authors and their affiliations should also be clearly shown, together with the abstract number. The latter is very useful to those seeking your poster, particularly at large meetings where the abstract number may not necessarily correspond to the poster board number. A photograph of the authors, particularly those presenting the poster, is a useful addition. It is an opportunity for you to promote your professional image, and it enables readers to recognize you in case they wish to make contact or ask questions. However, take care in choosing an appropriate photograph that portrays the image you wish. When you are not "manning" the poster, a poorly chosen photograph may prejudice the way in which the work is perceived (see chapter 2).

In terms of promotion and corporate identity, a logo of the organizations to which the authors are affiliated may also be added. However, the size and impact of this should not be inconsistent with the text size and the color of both the text and the background.

As a presenter you will want the poster to be noticeable and easily found. To this end, the title should be positioned as high as possible on the top of the poster, leaving only a small margin of 10 to 20 cm, so that it can be seen above the heads of viewers. However, a clear border should be allowed for in the layout design between the title and the body text. Letters for the title should be a thick style of text, a minimum of 25 mm high, but need be no larger than 45 mm. Whether dark print or color is used, a strong contrast to the background is required for reading from a distance. I would recommend use of some color in the title either as a background, a border, text, outlines, or drop shadows. Color works well when set in a large, heavy typeface, such as sans serif, and this may help to draw the attention of the audience to the poster.

The majority of posters produced by scientists without the assistance of graphic designers use capitals for the title and upper and lower case for the authors and for their affiliations. However, I would strongly recommend using a mix of capital and lower-case letters for titles as this is easier to read than all capitals, and it allows for the correct use of abbreviations such as ATPase, other curtailed words, for example amps, and terminology. Our eyes are, after all, accustomed to lower case with portions of the letters extended both below the lines and to different heights above the lines.

The wording used for the title of the poster should not be taken lightly.The text of the poster title and the authors' affiliations should be kept to a minimum. Careful consideration should be given to produce a form of words that is concise yet actively communicates the main part of the research findings. Titles such as "Studies of the involvement of both endo- and exotoxins in the pathogenesis of *Aeromonas hydrophila* infections" might be changed to a style that actually communicates some portion of the conclusion: "Endo- and exotoxins potentiate *Aeromonas hydrophila* infections."

The text of the poster title may be either centered or aligned left or right as befits the overall layout of the design. Where names of affiliations are wordy, they may be abbreviated in the title as they can be found in full in the published abstract. Similarly it is not necessary to give full addresses of the affiliations; just the name and minimal details of the location are sufficient. Where multiple affiliations are involved it is common practice to use a system of symbols to signify the relevant affiliation after the authors' names and to then list them. Although an author's middle initial may be omitted from the title, his or her first name should be given in full as this often helps facilitate direct dialogue between you and your audience.

8.1.2. Summary Section

Including an abstract in the title area, or elsewhere on the poster, is rarely necessary because its content is usually repeated in the text of the poster and is therefore redundant.

8.1.3. Introduction Section

The introduction should contain just enough text to justify your research and to present the aims of your study. The aims in particular should be stated as early as possible and should be clearly discernable. Benefit may be made from drawing the aims out of the text and listing and highlighting them by some suitable means, for example, bullet points, bold print, strong color, and so forth. Alternately, some presenters favor inclusion of a clear, separate statement of the objectives of the study.

8.1.4. Methods Section

The methods section should be concise. However, it is important to provide sufficient information to make your study credible and to demonstrate how the results were obtained. Generally, established techniques and equipment can be referred to by name, for example, polymerase chain reaction, scanning electron microscopy, and fast protein liquid chromatography, without going into the details of the procedure. Once mentioned, whenever feasible abbreviations should be subsequently used, for example, PCR, SEM, FPLC, to reduce the required text. However, avoid the temptation to invent acronyms or to generally overuse those already conventionally used, as the resulting text will become very disjointed and difficult to read. The use of diagrams, flow charts, lists, and photographs may be useful to convey which pieces of equipment and procedures were used. It may also be useful to cite methods used by others when appropriate.

8.1.5. Results Section

This section is particularly important as it presents the data in support of your research findings. However, raw data is rarely required, and only information relevant to your message need be presented. Make full use of graphic elements including clear tables and illustrations, for example, pie charts and photographs, to present the data (see chapter 7). *Remember, posters are a visual form of communication.* The use of text should be restricted to drawing out and explaining the relevance of significant points.

8.1.6. Conclusions Section

The conclusions drawn from your results should be clearly but concisely displayed. This is usually achieved by listing them and emphasizing them by some means such as the use of bold type, prominent color, bullet points, and so forth.

8.1.7. Discussion Section

The poster format is not ideally suited for discussion of results, as posters are by design a visual form and such deliberations will inevitably require a considerable commentary in the form of text. Discussion should therefore be limited and is often included with the results under a single heading of "Results and Discussion."

8.1.8. References Section

References are required if literature has been cited in the text. It is commonplace to list these in a smaller type size and in a less prominent position than other sections of the paper.

8.2. FLOW OF INFORMATION

The sections of your poster will need to be positioned in such a way that enables the viewer to easily follow your line of reasoning. The previously discussed format offers a logical sequence, but whatever format you choose, your flow of information should follow the basic principles of Western languages by reading from left to right first, then from top to bottom, and clockwise if wrapped around a graphic element or the circumference of the poster. Your poster layout should be specifically designed to be easily navigated, but you may feel on some occasions that numbering or alphabetically labeling the progression of the flow of information would be helpful. More visual indictors, such as large arrows or linkage bars, may also be useful within sections and can be incorporated into your overall design. Figures 8.1 and 8.2 show examples of the way in which the flow of information may be visually presented.

8.3. OVERALL DESIGN

It is important that the overall appearance of your poster gives a sense of harmony to all of the sections and to the elements contained within them. The separate sections must be easily discernable and capable of standing alone but simultaneously they must coalesce to form a single entity. With this in mind, you should take care to avoid displays consisting of a conglomerate of rectangular white paper sheets, as this will contain many inappropriate sharp angles and jagged edges that will distract the reader's eye. Small pieces mounted separately, or in a haphazard manner, will also serve to distract the viewer. Posters should be designed so:

- they communicate your research findings with immediacy and clarity;
- text, illustrations, and other graphic elements are organized into a cohesive whole for easy comprehension;
- visual elements predominate;
- the scientific messages are not masked;
- information flows smoothly and logically from one section to another;
- emphasis is put on those things that are of greater importance; and
- they contain no nonessential visual elements.

8.3.1. Designing Your Poster Layout

Before you attempt to make any layout for your poster, find out the exact dimensions of the display board that you will be using. I would suggest you divide your poster into two to six vertical strips. The exact number you choose will depend on the size and shape of the display board, but three or four are commonly used to good effect. Some variations in layout

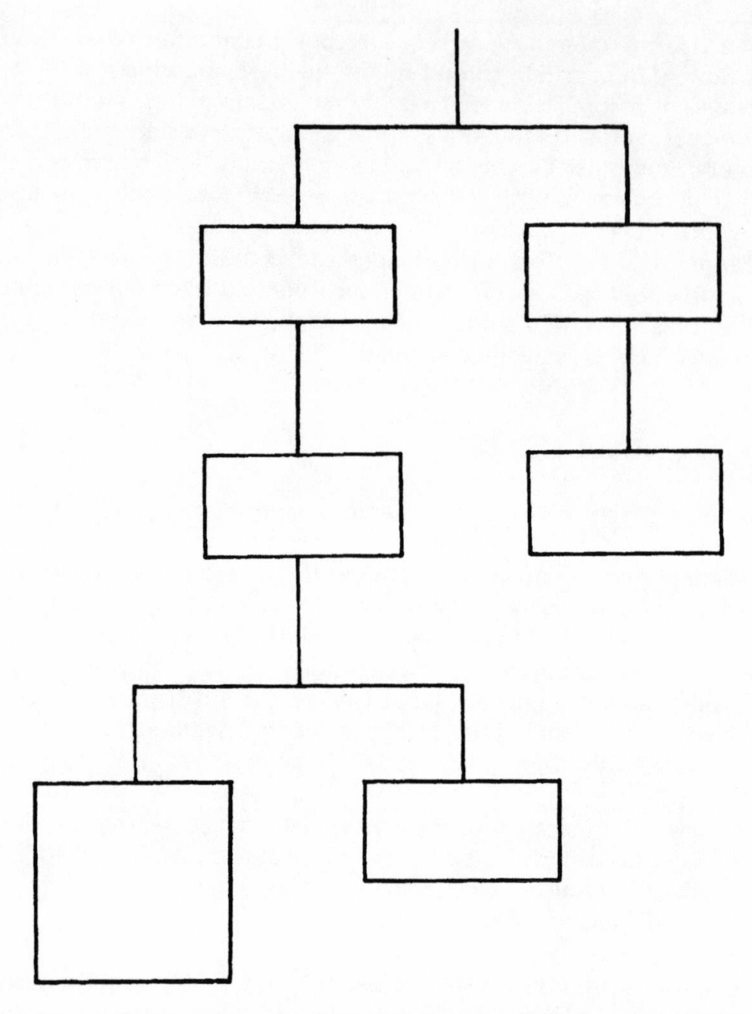

Figure 8.1. A skeleton scheme for directing information flow through two arms. This scheme enables text to be entered into the relevant boxes, leading the reader through the information contained in the two arms in a logical manner from top to bottom.

based on two or three strips are shown in Figure 8.3. Pieces of work that are closely related should be grouped together logically in one section or positioned near each other. To maintain clarity, subheadings or captions for figures and tables should be included in the same frame as the relevant text or illustration. Clear demarcation of sections and emphasis put on

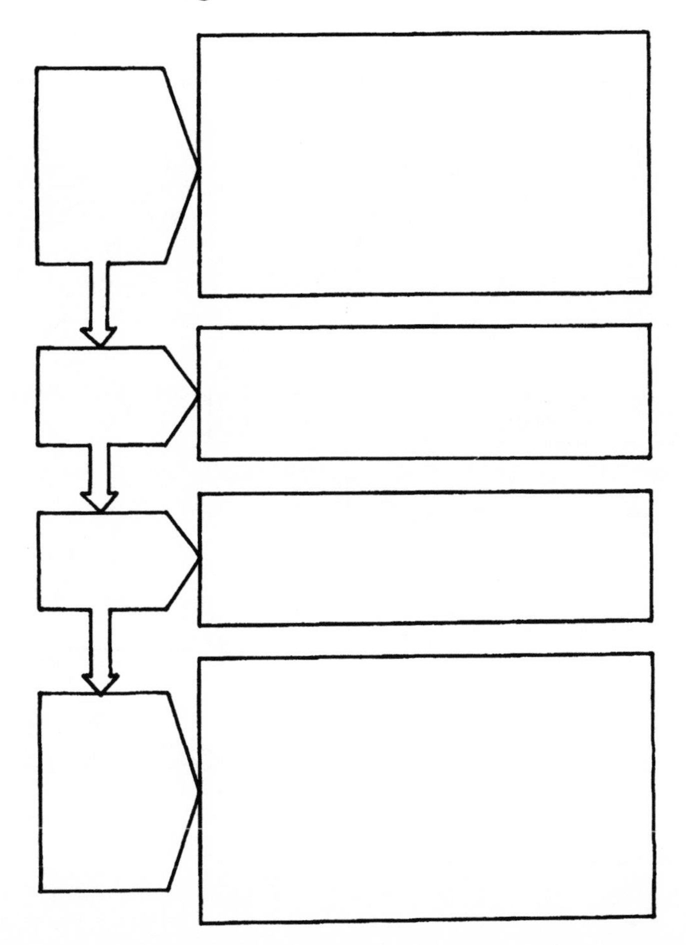

Figure 8.2. A skeleton scheme for directing information flow through a single dimension. This figure leads the viewer through text that may be entered into the relevant boxes, in a single direction from top to bottom.

important aspects can be accomplished by effective use of the arrangement of the various sections, the shape, the size, the color, and the text size, but all elements need to be carefully blended to produce a pleasing, unified effect.

8.3.1a. Creating a Scale Draft Layout. The following sequential actions outline the steps required to produce a scale draft layout of your poster.

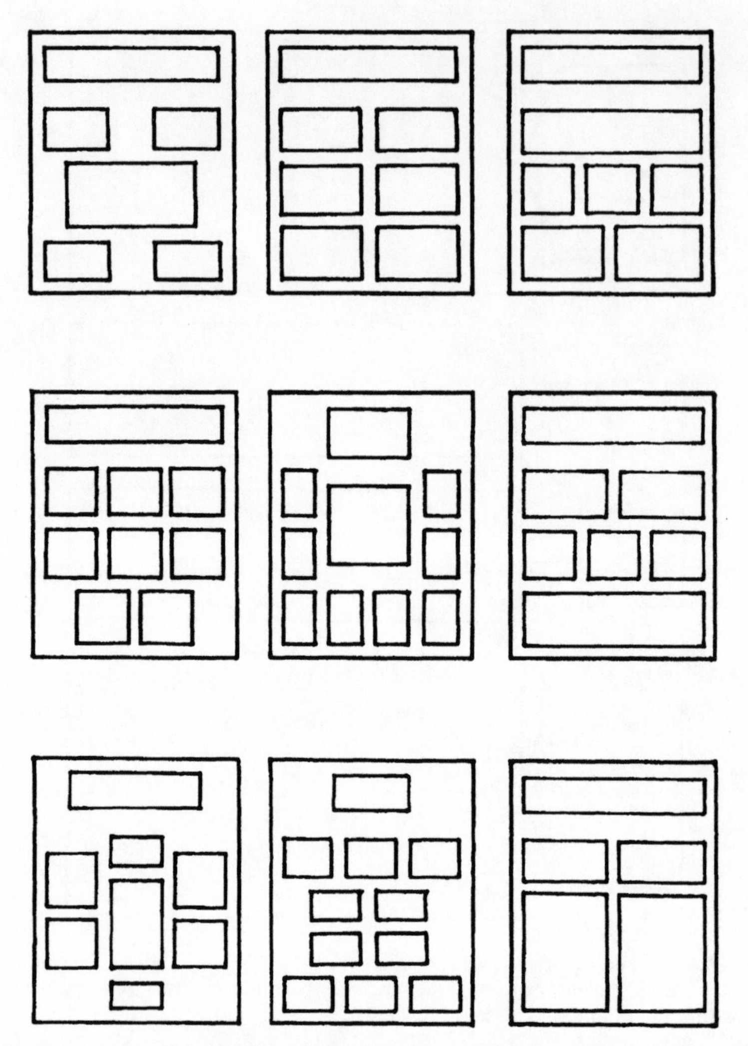

Figure 8.3. Some variations in layout based on demarcating the poster in to two or three vertical strips. These layouts demonstrate some frequently used poster formats. There are a wide number of variations that may be adopted. These examples show the importance of maintaining areas of clear space within the poster design.

1. Select a suitable scale of reduction to produce a convenient working size for a layout. This will be largely dependent on the final overall size of your poster. On a sheet of graph paper, measure the scaled-down outside dimensions of your poster and outline it.

2. Divide the rectangle vertically in thirds or in quarters. These divisions suggest possible columns to follow in the layout and, later, logical divisions for cutting and folding the finished poster. The number will probably be suggested by your rough sketch outlined during your initial planning stage.
3. Draw a horizontal line approximately one-eighth of the poster's height from the top for the title area.
4. Use paper cut to scale to represent graphic and textual elements. For instance, a number of pieces of paper representing your center graphic element and your introduction can be cut out. These paper elements can easily be moved around and various alternatives can be compared by taping the results temporarily and photocopying them. Place the pieces so they do not crowd the edges. Remember to avoid too many small pieces, which can give a "busy" appearance to the composition and blocks that are particularly large and overpowering. Experiment with the layout until you decide on one that allows the various components to fit the available area in a format conducive to your requirements.

The layout can be transferred to the actual poster by scaling up the dimensions.

8.3.2. Blank Space

Communication can be enhanced by the sizes and shapes of poster segments as well as by their positioning on the board. The blank space surrounding graphic and textual elements has a far greater role than merely separating the segments. Blank space is important as it dictates the "openness" of the design and when used effectively can communicate relationships among the parts. A good poster composition has a substantial amount of blank space, perhaps as much as half of the total area available.

8.3.3. Adding Flair and Panache to Your Poster

Adding flair and panache to your poster design will not only make an attractive attention-grabbing poster, if used appropriately it will ease reading and comprehension. Colored borders, bullet points, and blocks or bands of color can favorably add to the design while emphasising specific text or figures. Drop shadows of cut paper, chart tapes, or markers can make artwork appear to stand out from the surface. Three-dimensional mounting on pieces of cardboard or foam-core board dramatizes photographs and other important pictorial materials.

8.4. DESIGN EVALUATION

The following checklist should enable you to evaluate your design and to make sure that you have considered all the major aspects likely to affect the communicating power of your poster

- Is it the right size and shape for the display board?
- Is the title easily seen in a crowded room?
- Are the authors and their establishments easily discernable?
- Does it have visual impact?
- Is it pleasing to the eye?
- Does it have a clear visual central focus?
- Is the text easy to read?
- Is the poster easily navigated?
- Is the scientific message clear and succinct?
- Does it have an air of credibility and professionalism?

9

THE USE OF COLOR

The judicious use of color can transform a rather tired looking presentation into a vibrant, eye-catching display that even the most casual of readers will wish to read. However, there are definite pitfalls for the inexperienced presenter. The misuse of color in scientific posters may produce disastrous results beyond mere distraction from your scientific message. This chapter considers the use and misuse of color and gives guidance on when it can be used to its best advantage in the overall poster design.

When considering the use of color in a scientific poster it is important to remember the overall wider aims of the presentation, considered in chapter 2. It is important that your poster conveys your scientific message in a visual format, but it is vital that it does so in a controlled, professional manner that maintains the credibility of the scientific content, the scientist, and his affiliation.

Color possesses tremendous communicating power. Perhaps more than any other aspect of the poster design, color can have a devastating effect if it is not used wisely. Before considering specific areas in which color could be applied, it is prudent to reflect on the emotional responses different colors may evoke.

9.1. EMOTIONAL RESPONSES EVOKED BY COLORS

The color wheel shown in Figure 9.1 will serve as a guide in understanding the relationships between colors, the emotional responses they evoke, and will assist in the creation of a harmonious color scheme for the poster. Colors ranging from yellow-green to red evoke warm emotional responses. These colors are aggressive and eye-catching and tend to make objects appear closer to the reader. Conversely, colors ranging from green to violet are cool colors that evoke a soothing response that can, while making objects look clean and invigorating, appear to recede from the viewer. Colors that lie directly opposite each other on the wheel are complementary to each other. These will be particularly attention-grabbing when used in close proximity and will compete with each other to evoke a sense of excitement and vibrancy. Those colors that are close to each other on the wheel, such as shades of blue and purple, and shades of red and orange, when used in close proximity can compliment each other.

A most important use of color is to communicate emotion. Bright colors, such as shades of red, orange, and sky blue, communicate energy or optimism, whereas darker colors, such as shades of gray, navy blue, and maroon, communicate conservatism.

9.2. WHEN COLOR SHOULD BE USED

Color works best when used boldly over large areas or elements of the poster, for example, the background, large areas of text, or graphic elements. However, too many colors on one poster, or several small areas of color, add clutter and de-emphasize the scientific message. Its use should therefore be restricted to situations where it serves a positive design purpose, for instance by organizing blocks of text, or for emphasising key text. It is pointless using color for thin lines, or for small text, as it will be lost in the design.

Color should augment a well-designed poster; it should not be added in an attempt to paste over the cracks in a poor design. The design should allow the scientific message to be clear regardless of the colors chosen to amplify and organize them.

9.2.1. Beneficial Use of Color in Poster Design

When choosing which colors to incorporate, be guided by the impact and emotional response they evoke. Providing these are borne in mind, color may usefully be used as a background color, to highlight particular sections of text, to emphasize particular words, to highlight particular headings or subheadings, or to add impact to graphics.

9.2.1a. Background Color. The choice of background color is an important consideration as it will not only serve to create a unified poster, it will dictate the suitability of subsequent color use and set the tone of the pre-

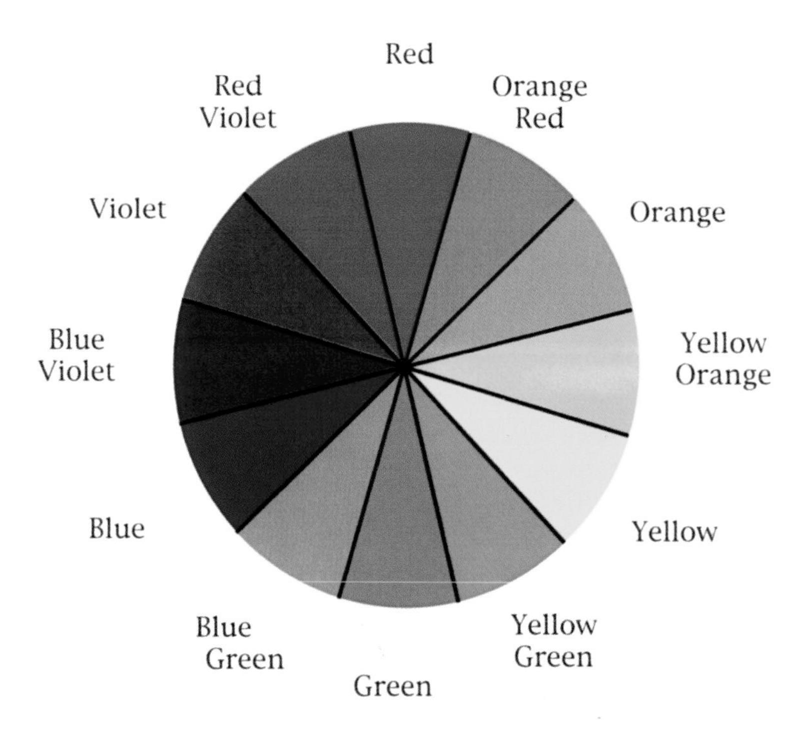

Figure 9.1. Example of a color wheel. Color wheels are useful as an aid to the judicious use of color within the poster.

sentation. Unless you are unfortunate enough for an adjacent poster to have an identical or very similar background color it will additionally serve to differentiate your poster from those nearby. The final choice of background color will need to take many things into consideration to achieve a satisfying balance of all the elements. Brilliant, intense colors should be avoided as although they will be very noticeable, they will draw attention away from your scientific content.

It is probable that subdued, neutral, or dark colors will be most often chosen as providing the most versatility. Subdued shades of blue, brown, earth tones, green, or gray are commonly effectively chosen.

It is probable that your poster will contain other color elements, whether in the form of highlighted text, photographs, or figures, so it is important to consider this right at the beginning to ensure that all elements may be conducive to producing a unified presentation. For example, if during your layout planning stage you have decided to include a central color photograph that consists of a predominately dark blue hue, it is probably ill-advised to use that as a background color.

It is common to use a very light color, for example, off-white, beige, or white, for the background paper for text because it works well. These colors conform to the expected norm, and with black or dark-colored ink they provide good contrast to ensure that the text can be comfortably read.

9.2.1b. Highlighting Sections of Text. Sections of text may be highlighted by means of a second background color or by the color of the text itself. In the same way that not all backgrounds need to be white, ink need not always be black. Combinations of dark, rich-colored text on paper tinted with complementary colors may provide an interesting alternative provided that the foreground/background contrast is maintained to permit easy reading. Care needs to be taken, however, as small serif typefaces set in color may be very difficult to read.

A different colored background to that of the main poster may be used to highlight and organize sections of text or data. Sections highlighted in the same color will appear related to each other, particularly if positioned in close proximity. Conversely, sections that are not highlighted, or highlighted in a different color, will give the impression of not being related. Highlighted sections will stand out from those elements not highlighted and will put emphasis on the text. These should therefore contain important aspects of your scientific message. Whole segments of the poster, such as methods or conclusions, can thus be effectively tied together by use of color highlighting. The use of colors that are close to each other on the color wheel, for example, blue/green and green, may be used as a code for different but related sections.

Alternately, or additionally, a color frame may be added to the section. Double or even triple mounting with contrasting colors can provide an eye-catching frame to the section of text, of data, or the graphic element. Shadowing effects can be easily produced by offsetting the mount, but as for shadowing letters this should be to the left rather than to the right.

9.2.1c. Highlighting Photographs. The techniques described previously to highlight sections of text can generally be applied to highlight photographs. However, the effect that the background color will have on the contrast and the colors of the photograph will need to be considered. In general, dark photos will look darker on a light background and lighter on a dark background; color photos will look more colorful when displayed on a neutral background, such as medium gray, and less colorful on white.

9.2.1d. Emphasizing Particular Words. Key words within the text may be emphasized by increasing their size, making them bold, and using a color that is complementary to the background. Choosing a color such as red will evoke a sense of excitement and will stand out and be noticed by the reader. Alternately, you may wish to reverse the colors, in this example using a red background. However, take care as in general, text set in color is significantly harder to read than text set in black, unless it is made larger and bolder.

9.2.1e. Highlighting Headings or Subheadings. The importance of size, typeface, and density on the impact of headings and subheadings was considered in chapter 5. Headings set in color are usually harder to read than those set in black. Color may be used, however, to organize the text by augmenting the separation of various sections, clearly indicating that a particular section is a separate topic. If used, care must be taken to choose colors that preserve the contrast between the text and the background and that stand out from the main body of text.

9.2.1f. Adding Impact to Graphics. Color may be particularly useful in adding impact to graphic elements of the poster. Columns in tables, lines in graphs, bars in bar charts, segments of pie charts, and the like may all benefit greatly by use of the discriminating power of different colors. Complementary, bright, vibrant colors, for example, red, blue, and green, should be chosen to maximum benefit. Color brings verve to the conveying of scientific information through graphic elements, in a way that is not possible using shades of gray, parallel lines, and cross-hatching.

Color can help organize the information within graphic elements.

Colored backgrounds can be used in place of horizontal and vertical lines or blank spaces to organize a table. This has the effect of grouping information and provides the reader with a visual guide as to the direction that the information should be read.

9.2.2. When Use of Color is Not Beneficial in Poster Design

To be beneficial color must not be scattered. It needs to be used over relatively large areas, such as background for title blocks and large sections of text, but not for small sections of text, small subheadings, or text set in the same size as the body copy.

10

PRACTICAL ASPECTS OF POSTER CONSTRUCTION

There comes a time when the outcome of the imagination and flair of the poster designer has to be fitted into a professional, workable form that is accurately constructed. Construction of the poster requires patience, accuracy, and some degree of draftsmanship.

10.1. CONSTRUCTION TECHNIQUES

Final construction of the poster can take place once the scale plan has been produced; the text has been written; the photographs have been prepared, having been enlarged or reduced as required; and the graphic elements have been made ready. *You should assemble the individual elements as indicated on the scale model layout plan.* Using the layout as a guide, mounting boards may be cut to size, and layout measurements can be easily transferred in soft pencil and erased after the poster elements have been affixed.

The means by which the presentation is to be assembled on the display board should have been decided as part of the initial planning stage. Having constructed each segment separately, you have the option to either position each one individually on the display board at the meeting, or position each one in their final positions onto a backing sheet. Whichever technique is adopted, the construction of the individual segments is a prerequisite.

Sections of text should be printed, preferably by laser printer, on paper or thin card that is durable and will not wrinkle once it is glued into position. If the segment is to be framed, the section of text, photograph, or other graphic elements should be glued by liquid or spray-mounting adhesive or stuck on with double-sided tape to the chosen backing card. It should soon become apparent that the construction stage is far from onerous. The careful planning and effective designing processes carried out initially should have predicted any potential problems and enabled you to have ironed them out well before attempting final construction. However, it is prudent to lay out all of your segments in position before final attachment to backing materials so a final review can be conducted. If necessary fine tune the positioning, but you should not be tempted to change anything too radically, as this would then have an impact on the design of the whole poster. If it becomes necessary to adjust the size of some element to make it fit, ensure that the size increase or decrease, which should be minimal if your planning was conducted adequately, does not influence the overall design. Graphic elements may be reduced or enlarged by making use of a photocopier but *never alter the size of a section of body text*, as this will have a marked effect on the unity of the design.

10.1.1. General Tips on Construction

Construction techniques usually involve drawing, measuring, cutting, and gluing. The following are some general tips that may assist you in conducting your poster construction:

- Be sure your cutting tools are sharp when you begin to cut paper or cardboard.
- Ensure your cutting surface is suitable, ideally use a purpose-built cutting board.
- Avoid spraying too much adhesive in an enclosed area; you may develop a headache or more serious reactions.
- To keep edges straight, hold a metal straightedge against your marked matting, and for cutting long edges ask someone to help hold the straightedge.
- Measure carefully before making a cut. Even small variations in frame widths or paper size can look magnified on a poster display.

10.2. BACKING MATERIAL

Color, depth, quality, and texture of backing materials can all contribute positively to the physical appearance and the communicating power of the poster. The same features, however, if chosen imprudently, can also serve as serious distractions from the scientific message.

The impact of the color of the background material is as much a function of the surface on which it is printed as the ink used to reproduce it. The use of color is considered in detail in chapter 9. Posters are frequently mounted on colored matte-finished cardboard that is stiff enough to carry the poster segments but sufficiently flexible to enable it to be easily rolled or otherwise packed for transportation (see chapter 12). However, you may be able to utilize a variety of materials, such as cloth, netting, cork, or others that are available from home decorating stores, art supply stores, and stationary shops. Many potential poster backing materials have a textured quality that if chosen wisely may be attractive without distracting from the scientific content. If you choose a less conventional backing material it is prudent to keep it within the theme of the poster content; for example, if presenting research concerning fish farming, a netted or matted background material, in a "natural" color such as light brown or green, may be effective. Good judgment can increase the durability of the poster and the ease of construction.

To make your poster more durable, you may wish to have the various segments or the entire poster laminated. This should be approached with caution as a flat finish can blur your text, and a glossy, laminated finish can create an uncomfortable glare to the eyes in a brightly lit exhibition hall. The latter also inhibits the use of flashbulbs in photographing your poster, a practice that is becoming increasingly popular with delegates.

Whatever you choose as the backing material for your poster, you must remember that it will unite the various segments and have considerable impact upon the overall appearance. It must be congruent with the theme and style of your poster and must not distract from the poster content.

10.3. PAPER CHARACTERISTICS

Segments of poster text are commonly printed on A4-sized, white photocopy paper. This practice gives reasonable results but the paper is not ideal, as certain inks bleed, and it has a tendency to wrinkle or tear when pasted to the support card or other backing material. There are many different papers that may be used when printing the poster elements. Paper comes in a variety of qualities, differing in weights, textures, and colors.

10.3.1. Paper Quality

The quality of paper depends on both its raw materials and its method of production. It is not the purpose of this book to consider the details of paper manufacture, but I would refer those interested to the volumes cited at the end of this volume. However, the quality and characteristics of paper may have a marked effect on its suitability for use in the poster construction. Paper made completely of cotton as its source of fiber are very high quality, but the raw material is becoming increasingly expensive. It is therefore common for paper to be made of a mixture based on a reduced cotton content and of plant fibers of a similar quality and origin, for example, flax, hemp, and jute. Paper of anything more than 50% cotton in composition is usually of a fair quality.

10.3.2. Paper Weight

The weight of a paper is expressed either in grams per square meter or in the old Imperial system of pounds weight. The term pounds refers to the weight of 500 sheets, referred to as a ream, of the paper in Imperial sheets (30" × 22"). A description of a particular paper as 90 lb or 300 lb does not refer to a specific quality of the sheet, therefore, but is a general indication of its density and thickness. The heavier a paper is the more stable it is. The use of different paper weights is relative to their end uses. An imposing letterhead, for instance, such as may be used for the poster title, will look better on heavier and thicker paper than the main text.

10.3.3. Paper Surfaces

A variety of manufacturing processes can bring about papers with differing surface characteristics that may be useful in poster construction. "Laid paper," for instance, contains a pattern of lines that is visible when held to the light or when lightly drawn on, whereas paper produced on a wove mold has a much less distinct pattern and is a more smooth-surfaced paper. "Hot pressed" paper is even more smooth in texture and excellent for drawing on using pen and ink. Exceptionally smooth paper surfaces may be produced by a process termed "calendering." There are, in addition, a number of other finishing processes responsible for surface effects that result in decorative textures, giving the paper, for example, a woven cloth appearance, which may have a place in your poster design.

10.3.4. Types of Paper

There are many different types of paper available, many of which are sold in pad form for convenience. These include layout paper, tracing paper, drawing or cartridge paper, line paper, surface paper, quadrille, graph and technical papers, cover paper, and marker paper. The various types differ in their quality, thickness, and use. Marker paper in particular may be of use in poster construction. This is a white paper that frequently has a very smooth surface. Typically, it is 70 gm^2 in weight, and has a slight degree of transparency. It is specially designed for use with marker pens and felt-tips containing either water-based or spirit-based inks. These inks are inclined to bleed on ordinary paper, and produce feathery edges rather than crisp edges along lines. Marker paper is bleed-proof. Most marker paper is stark white, such that the surface appears to have been coated, making it exceptionally smooth. Although intended specifically for use with markers, marker paper obviously lends itself to use with most types of pens.

10.4. DRAWING AIDS

No matter how proficient you are with computers, for the amateur poster producer it is most likely that some degree of manual drawing of graphic elements, even if it is only straight lines for underlining or boxing text, will be a necessity. A broad selection of drawing aids and instruments are available to assist, and a selection are considered in the following section.

10.4.1. Drawing Boards

A firm, flat working surface is important for virtually all poster construction procedures. Drawing boards are available in a variety of materials including plastic and wood, and portable, desktop, and free-standing models are available.

10.4.2. Rulers

A ruler is a combination of straightedge and measuring scale. A typical ruler is flat on one side, and on the other both long edges are normally beveled. Used with the bevelled side uppermost, the ruling edge is flat against the drawing surface. Used with the bevelled side down, the edge is lifted clear of the drawing surface and lines may be drawn along it with a pen or a marker. If the ruler is not used this way with pens and markers, ink tends to flow under the edge of the ruler as the pen is drawn along, and a blotted, smudged, and inaccurate line results.

Clear plastic rulers have the advantage, over wooden and metal ones, of transparency, which can be very useful when pasting, or mounting segments onto the backing material when the position of other elements in the design is relevant. Particularly useful are clear plastic rules with a series of evenly spaced parallel lines scored lengthwise along the center of the ruler. These may be positioned against an existing line or line of text, for example, and a reasonably accurate line can be constructed at a set distance parallel to it. This may be especially useful in cross-hatching a particular area of a graph or a pie chart, for example. However, the edges of plastic and wooden rulers are prone to damage unless properly maintained. *Never use these types of rulers as a straightedge guide for a cutting blade.* The edges of metal rulers are not so easily damaged.

10.4.3. Straightedges

A straightedge is usually not noticeably different from a ruler. It is a reinforced or specially fitted edge that ensures an accurate and unwavering line. A metal strip fitted into a wooden or plastic ruler, or a steel strip set into an aluminium ruler, is a most reliable form of straightedge. The metal strip frequently projects slightly so that the ruling edge is lifted cleanly from the paper, making it suitable for use with pens or markers. Of particular interest to those constructing a poster are cutting rulers. These have a shaped section designed to protect the fingers of the hand steadying the ruler while a knife can be drawn along its edge.

10.4.4. Drawing Devices

There are many devices that have been designed to improve, or accelerate, the drawing process. Several are concerned with accuracy, but others are intended just to make drawing easier, or to simplify some otherwise difficult operation. Those readily available include the following:

- Triangles. These are drawing aids for quick and accurate drawing of angles. Parallel lines may be constructed quickly using a triangle and a ruler, each line being constructed at a right angle to a horizontal or vertical guideline.
- Protractors. These are used for measuring or constructing angles, are in the form either of a semicircle or a full circle, marked off around the edge in degrees, thus representing either 180° or a full 360°. These are an essential tool when constructing pie charts.
- Pantographs. These are used for copying a drawing to a larger or smaller scale. These consist of four rods pivoted at their intersections to form a folding, extending lattice. For use, one end of the pantograph is fixed, a pen or pencil is placed at the end of one of the extending arms, and another is placed at the junction of the other two. When one of the pens or pencils is moved, the other re-

peats the movements, but on a different scale. These may be useful to the poster producer, in the absence of a photocopier that is capable of reducing or enlarging images, or where the image quality of a photocopy is not considered adequate.

- Templates. These offer a very wide range of shapes and sizes intended to assist drawing in both specific and general applications. They are in effect shapes that you draw around and are usually in the form of a thin plastic sheet with cutout shapes in a variety of sizes. Arrows, circles, ellipses, and lettering templates have general uses and may be of interest to many scientific poster designers.
- Curves. These are templates composed of several changing and graceful curves that offer fixed shapes that can be repeated, or at least rendered accurately, without the need for elaborate construction.
- Flexible curves. These are in the form of a strip constructed around a lead core, with a lip to prevent ink from flooding under the curve that can be bent to virtually any desired shape. It holds its shape while being drawn around, but can then be reformed to another shape if desired.
- Perspective grids. These are useful in rapidly constructing a perspective drawing. They are already marked with a detailed grid that follows the lines of perspective, and may be placed beneath the drawing, if the paper is semitransparent.

Other more commonly recognized aids include dividers and compasses, which are available in a variety of forms, that perform tasks pertinent to measuring distances, angles, constructing shapes, and describing circles.

10.4.5. Erasers

When a mistake is made on a drawing, the right eraser can often correct it. They are of course also useful for removing guidelines from a piece of artwork. Erasers are made of either soft rubber or plastic for use on pencil. Special erasers are made for removing ink, although these may not be completely successful, and have a tendency to erode or tear the paper.

10.5. ADHESIVES

Adhesives are necessary for the process of pasting up the poster, and a variety of alternatives are available.

10.5.1. Adhesive Tapes

There are a wide selection of different adhesive tapes available, including clear plastic tapes, colored or printed tapes, paper-backed tapes, gummed tapes, insulating tapes, and double-sided tapes. However, with the exception of the double-sided tapes, none of these are suitable for discreet final mounting. Double-sided tape, as its name suggests, carries an adhesive on both surfaces. Typically it might be used to mount photographs or other poster elements flat onto a backing board. Its advantages are that it is quick, clean, and cannot be seen. Materials mounted with it, however, tend to cast a slight shadow, and this may deter you from using it to paste-up your various poster elements.

For applications that require a stronger bond, or that have to carry weight, double-coated foam tape is more suitable than double-sided tape. This is a thin strip of foam with a powerful adhesive on each side. Double-coated foam tape also comes in the form of individual mounting squares. For use, an appropriate length of the tape is cut, positioned, and stuck down on the adhesive lower side. The top adhesive is then revealed by peeling back the protective coating. Foam tape usually takes a firm grip as soon as its surface touches an object, so there is little scope for repositioning. It causes a slight shadow to be cast, but it is excellent for mounting quite large display items onto backing material.

10.5.2. Aerosol Adhesives

Aerosol adhesives are useful when pasting-up work because they efficiently lay an even coat of glue over large areas, such as the whole of the reverse side of the poster element to be pasted up, insuring a firm, flat bond. A spray is ideal for administering sufficient adhesive to be effective without the use of such a quantity of glue as would ooze during mounting, and would require a further cleaning operation. However, care should be taken as some aerosol adhesives can stain or discolor artwork. Most aerosol glues allow for repositioning and are either permanently removable or have a prolonged setting time during which the work can be lifted and repositioned. For temporary bonding the glue is applied to one surface only, but if you require a more permanent and secure bond, both surfaces must be coated after which repositioning becomes difficult.

Aerosol adhesives are generally available in three different grades:

- a plain mounting quality intended for paste-up work and general light-duty mounting.
- a photomounting quality, which is a heavier duty adhesive capable of a more permanent bond.

- a heavier quality still for display mounting when the objects to be adhered are of a large size or are made of a heavy material such as illustration board.

Although suitable in most normal circumstances, I would not recommend using basic mounting glue supplied in aerosol form for pasting-up posters that may be stored intact for repeated use at a later date. Over time, these tend to lose their adhesive qualities and pasted-up work may well peel off. In these circumstances a photomounting or a more permanent type of mounting glue should be used.

10.5.3. Rubber Cements and Glues

Rubber cement or glue is the more traditional approach to pasting-up and mounting poster elements. Other paper glues, such as gum arabic, are rarely used now. Rubber and latex cements must be applied with a brush or spatula to the back of items being glued, and they can be rather messy to use. Experience in their use is required to avoid bubbles and blobs appearing in the surface of the artwork or excessive seepage to occur from the edges when it is pressed down.

Rubber cements dry comparatively slowly, allowing a good period of time during which the work can be repositioned. Although suitable for most poster applications, rubber cements cannot always be relied on to form completely permanent bonds.

10.5.4. Adhesive Sticks and Pens

Adhesive sticks and pens are available to cope with small jobs and applications in which particularly accurate use of glue is required. The paste is rubbed onto the back of the artwork from the stick, and as the stick wears down it is further extended by pushing or twisting the base of its container. Glue pens are available with different shaped heads through which they deliver glue allowing for accurate application.

10.6. CUTTING EQUIPMENT

Knives and scissors are common items that are suited to the small cutting operations likely to be required during poster construction. Some of these have special applications and can cut curves or special shapes easily. For larger straight cuts, paper cutters, paper trimmers, or guillotines should be used.

10.6.1. Cutting Mats

The best results from cutting with art knives and other small-bladed instruments are obtained when they are used in conjunction with professional-quality cutting mats. These have a non-slip surface that holds the artwork firmly as the knife cuts and are generally overprinted lightly with a grid to assist positioning. The surface of a cutting mat is usually constructed of a special material that is "self-healing," closing up after being cut so as not to leave a score mark, which might misguide the blade when it next follows a similar path.

10.6.2. Art or Craft Knives

Art or craft knives are ideally suited to very fine cutting and trimming operations. There is a wide choice of blade shapes available to suit almost any requirements you are likely to have during poster construction.

10.6.3. Scissors

Scissors need to be extremely sharp to cut paper well. For cutting perfectly straight lines and for cutting cardboard, scissors are not particularly useful as they are inclined to wander, particularly when cutting heavyweight materials. However, some high-quality scissors are available that give above average performance. Scissors with very long blades tend to be best for straight cuts. Needlework or surgical scissors are recommended for delicate and accurate cutting.

10.7. INSTANT LETTERING

In constructing some of your poster elements, such as the title or the main headings, you may require letter styles or sizes that are not available to you on your computer, word processor, or typewriter. In these circumstances instant lettering systems are useful and available in a variety of sizes, content, and style. Lettering styles include white, black, other colors, metallic finishes, and a large selection of shading, lines, borders, decorative patterns, and instant illustrative material in the form of dry transfer or self-adhesive sheets. Many of the dry transfer or self-adhesive sheet products other than lettering are, in fact, aimed at specific applications including science and engineering.

10.7.1. Dry-Transfer Sheets

The principle of dry-transfer lettering is relatively simple. A typeface is printed in reverse on the back of a clear plastic sheet, using flexible inks, so the bond between the printing and the plastic sheet is weak. The printed side of the sheet is protected by a backing sheet of waxy or silicone-coated paper that does not stick to it, and that protects the delicate type from accidental damage.

The quality of the plastic sheet and the properties of the inks used for dry-transfer products vary between suppliers, but the best inks do not easily crack or break as they are applied, and are heat resistant. The plastic carrying sheets should be of an appropriate thickness, and the best ones are made of polyester, which is able to withstand the pressures of burnishing well.

10.7.1a. Application of Dry-Transfer Products. To obtain the best results using dry transfer products, a specific method of application is recommended. The following procedure is recommended for consistently good results:

1. Apply a light guideline to the artwork, or place sight marks along its edge, so a ruler can be used as an aid to positioning.
2. Remove the protective backing from the transfer sheet. Hold it so the type or design is seen as it will appear, and place it over the artwork and gently maneuver it into the correct position.
3. Smoothly and repeatedly draw a burnisher, or any convenient instrument such as a pen or pencil which may be used as a burnisher, over the chosen letter, numeral, or design element, making sure that every part of it is well rubbed down.
4. Lift the plastic carrying sheet from the artwork with a *gentle peeling* action, leaving the print in position on the surface of the artwork. It is possible to see that the print transferred before removal of the sheet, because the separation of the printed ink from its carrying sheet causes it to appear gray when viewed through the sheet.
5. Make any corrections that are necessary. Dry-transfer materials can be removed from smooth, nonabsorbent surfaces with tape, and from other surfaces with a hard eraser specially made for the purpose.
6. When all of the transfers have been applied, place the silicone backing sheet over the applied transfers and burnish the whole area again to ensure a firm bond between the transfer and the artwork.
7. Erase any guidelines carefully, so not to disturb or damage the transfers,and clear any remaining pieces of eraser from the surface by blowing or by removing with a very soft cloth.
8. Apply an aerosol spray fixative, as recommend by the manufacturer of the transfers, to secure a long-lasting protective surface.

10.7.2. Self-Adhesive Film

Self-adhesive film is an alternative to the dry-transfer process. It consists of a thin plastic sheet on the front of which the letters are located, that is bonded to a waxy or silicone-coated backing sheet from which it is peeled before application. Self-adhesive film is applied as cut pieces that overlap their intended area, and these are trimmed with the aid of a knife. The soft tack of the adhesive allows them to be removed easily for repositioning and permits excess trimmings to be removed without difficulty.

Self-adhesive films are particularly favored for larger illustrative elements. When self-adhesive films are placed in their final position, I would recommended that you burnish them all over to improve their hold on the artwork.

Large self-adhesive letters are also available. These are die-cut into paper, or more often into vinyl sheets, and are mounted on a protective backing piece. These large letters and numerals are sufficiently self-supporting to be handled on their own and can be lifted from the sheet by slipping the tip of a knife blade under the edge. They can be carried and positioned in the same manner, on the end of a knife or a special handling spatula, and once in place should be firmly rubbed down to ensure that all of the air is removed from under them and that they are firmly fixed to the backing material. Lettering of this type may be used on large-scale artwork, such as the poster title, to produce a professional finish.

11

THE USE OF SUPPLEMENTARY INFORMATION

In this chapter we consider the use of supplementary information in a scientific poster presentation. It is important, however, that the scientific information being visually projected on the poster is complete itself and does not rely on such supplementary information to convey the scientific message. This is communicated primarily through the poster itself and the abstract published in the proceedings of the meeting. There are additional materials, however, that can be made available at the meeting that will help the viewer record your findings and make your message more lastingly noticeable. Your credibility as a scientist may also be enhanced by an obvious display of appreciation of the needs of scientists in the field.

11.1. CONTACT INFORMATION

Every piece of useful or informative material offered to your audience should be attributed to you. You should take every opportunity to raise your own profile and encourage further contact.

11.1.1. Business Cards

Business cards provide a very convenient means of disseminating your contact details. If you possess them you should ensure you have sufficient quantities before attending the meeting. They should be made available by leaving a supply in a box clearly labeled near your poster. They may also be placed in a pocket that is attached to the board, adjacent to your poster. It is a good idea to write on the back the name and date of the meeting, together with your poster title to act as a reminder to those taking them.

11.1.2. Contact Envelopes for Communication

A small (A4) letter box, attached to the poster board and positioned directly adjacent to the poster, for requests for additional information is useful. Such boxes can be easily prepared from boxes used for A4 photocopy reams or similar materials. A slot can be cut in which communications can be dropped. It is useful to provide a supply of index cards and a pen, usually attached to the board with a piece of string, for addresses and comments. The purpose of the box should be clearly stated on the front. Often, business cards with notes on the back are left in such boxes. During the course of the meeting the box should of course be emptied, and it is prudent to keep a record of the details of those communicating, together with the message in a book or file, as small notes are prone to be mislaid among the plethora of papers one tends to accumulate at scientific meetings. Clearly any contacts obtained by this route should be followed up as soon as possible, ideally face-to-face at the meeting. If this is not possible, however, you should follow up contacts immediately upon return from the meeting. Take the trouble to send any information requested, but establish personal contact whenever possible in doing so.

11.2. SUPPLEMENTARY INFORMATION

Any supplementary information that will help in conveying your scientific findings, communicating your views, or projecting yourself as a credible scientist, which you could easily make available, has a potentially valuable role in augmenting your poster presentation.

11.2.1. Important Information Sheets

Information sheets containing important aspects of your presentation are a useful supplement to the poster itself. These can be placed in an open envelope or pocket attached to the board adjacent to the poster. Such sheets again should be headed by the authors' names, addresses, and contact numbers. These may prove valuable to the viewers after they have left the meeting. Take care in producing such sheets, however, as they may well supplement the written abstract published in the proceedings of the meeting, and need to be of similar quality.

11.2.2. List of References

It is probable that a list of references will be integrated into your poster design. However, viewers who are particularly interested in your field may wish to follow the derivation of some or all of your work. Members of your audience may wish to obtain copies of your citations, and it will be most helpful to them if you supplement your presentation with sheets listing your references. These sheets should be headed by the poster authors' names, addresses, and other contact details including e-mail addresses and phone and fax numbers. These viewers probably represent one of the most interested groups, and you will wish to reinforce your details as the source of the citations. The envelope, box, or pocket should be clearly labeled and refilled as necessary during the presentation.

11.2.3. Reprints

If the work you are presenting has been published prior to the meeting, a sleeve attached to the poster containing reprints is useful in providing further details. The fact that the work being presented has also been peer reviewed will add credibility to your findings and to your presentation.

11.2.4. Copies of Your Poster

It is becoming increasingly common for presenters to provide reduced copies of part or the whole of their poster for viewers to refer to at a later time. These may be produced either through computer printouts or by reducing elements on a photocopier and are frequently made available in color.

12

TRANSPORTING AND ASSEMBLING THE POSTER

In this chapter we consider the means by which you can transport your poster to the meeting venue, with confidence that it will be in the same condition as when you left.

12.1. TRANSPORTING THE POSTER

The simple act of getting the poster to the meeting can be fraught with difficulties. Anyone who has tried to carry any large, flat object will vouch for the fact that the wind-trapping capacity of such articles is quite remarkable.You will need to be able to conveniently pack and transport your poster to the meeting in a manner that ensures it arrives in a state suitable for display. It is therefore most improbable that you will attempt to transport your poster whole and flat.

12.1.1. Rolling

One solution is to roll your complete poster and either secure it with elastic bands or with adhesive tape attached to the back. If the backing material is firm enough this may then be simply wrapped in brown wrapping paper to avoid marking. It is undoubtedly preferable, however, to place the rolled poster into a ridged cardboard tube, or a specially built plastic carrying tube, which are readily available in a variety of sizes from stationery stores or artist supply outlets. A word of caution, however. It is not advisable to roll a poster that has sections or graphic elements already glued to it. All too often the poster, upon unrolling at the venue, is in a sorry state — sections peeling, curled, and wrinkled! If the rolling option is taken, I would suggest that the backing sheet is transported thus, with the sections transported flat, separately. The backing sheet can have the position of the sections marked on it in pencil, and you can finally reconstruct the poster at the venue.

12.1.2. Folding

It is not feasible to simply fold a completed poster and expect it to unfold in a condition that is suitable for display. The folds will be most distracting, and will preclude the poster from lying flat on the display board. Some consider it an option to construct columns or sections of material, cut them into smaller segments, and use duct tape or strapping tape to hinge the back so you can fold them for packing and simply unfold them to place the connected pieces on the board. This technique is not especially good, however, as inevitably the site of the joint has a gap through which the tape shows in a distracting manner.

12.1.3. Segmenting

Segmenting the mounted poster into three or four medium-sized strips is a popular means of providing a manageable form in which the poster can be transported. The sections can be positioned close to each other, or simply joined edge-to-edge, on the display board. However, the use of shiny transparent or colored adhesive tape should be strongly avoided as it can look unsightly, reflect the light, cannot be easily removed, and does not allow readjustments to be made.

12.2. ASSEMBLING THE POSTER

Assembly is one of the most important steps in the overall process of poster presentation and is also one of the most often neglected. It is vital for your poster design that components are affixed in exactly their allotted position on the display board and in relation with each other as dictated by your scale drawing, and that the whole composition is placed in the optimum position on the board. All too frequently, what could have been an effective display is ruined by a rushed, shoddy assembly. The human eye can detect even the slightest jagged line, slope, or uneven spacing and these errors can accumulate to destroy the unity of your poster. Worse still, they are most distracting to the viewer, who will not be encouraged to proceed to give your poster the attention it should deserve. You will have spent too many days, weeks, and possibly even months planning and compiling your poster to let such events take place. They are easily avoided. *Above all, you must allow sufficient time in your meeting itinerary to accurately assemble your poster.* To make optimum use of the time available, you should be sure you know, before leaving for the meeting,

- your poster board number,
- the times that you have access to the room to mount your poster,
- the deadline for posters to be mounted,
- the day and time of your poster presentation, and
- the deadline for removal of your poster.

The manner in which you are able to attach your poster to the display board will largely depend on the type of board used. The meeting organizers should be able to supply information on the type of board and on the preferred method of attachment. The options for attachment generally include pins, Velcro, magnets, and double-sided tape. Any of these will enable you to successfully attach your poster. With this information you will be able to commence setting up the display at the earliest opportunity.

The time involved should not be underestimated. The sooner you set up your display the better. This will enable you to capitalize on the maximum exposure for your work and free your time for other matters, such as visiting other posters and making new contacts.

To avoid the presentational pitfalls that may occur during assembly of your poster, you should ensure that

- columns and rows are aligned precisely,
- the components are horizontally and vertically aligned to the display board,
- spaces between components are equal,
- components are firmly attached to the display board,
- the title is high on the board,
- the main body of the poster is between eye and waist height, and
- the poster has an equal border on each side of the display board.

To achieve such precision you will find a tape measure, ruler, and right-angled triangle essential, and I would recommend that you ensure you take these with you to the meeting. Depending on the content of your display, a piece of string that you could use as a plumb line and a small spirit level, such as used by photographers in making sure their cameras are level, may also be useful. One of the simplest ways to make sure columns are straight is to accurately measure where the edge of the column is located and pin a piece of string along the entire length of the poster board. The string can then be used as a guide for positioning components and moved when completed. Similarly, string drawn diagonally from the corners of the display board will form a cross indicating the precise center point of the board, which, depending on your poster construction, may be useful to ascertain.

If your poster is in the form of a single sheet that has been transported rolled, most of the potential pitfalls will be bypassed; however, even then it is vital to ensure that it is placed on the display board optimally. Be sure that the title is high on the board and that the main body of the poster is between eye and waist height, and that it is positioned horizontally, with an equal border on each side.

12.2.1. "Assembly and Repair Kit"

No matter how carefully you have planned, transported, and assembled your poster, the unexpected tear, mark, crease, or other mishap may occur. I would recommend you prepare for such eventualities by taking to the meeting with you a "poster assembly and repair kit" for use in mounting the poster and for on-the-spot repairs. I have listed some of the items you could usefully include in the kit. These should enable you to repair any damage and to make small corrections to the content that for whatever reason were not recognized before. I would also recommend including a variety of different means of attaching your poster, as it is not unknown for last-minute changes in the type of display board to occur. Contents of the poster repair kit should include the following: pins, adhesive, Velcro tape, pencil, pens (various colors), an eraser, double-sided tape, marker pen, ruler, business cards, supplementary information sheets, containers/envelopes for information sheets, correction fluid, scissors, a pencil sharpener, an art knife, a tape measure, string, and a right-angled triangle.

12.2.2. Supplementary Components

When you have assembled the poster, you should attach the envelopes or small boxes that will contain your supplementary information to the display board. You will have to ensure that there is a sufficient supply of these during the course of the meeting. I would suggest, however, that it is prudent to reserve a suitable supply for distribution at the timetabled poster session.

13

PRESENTATION AT THE "POSTER ROUND"

Most national and international scientific meetings now include a "poster round" session. During these sessions the author is expected to be present at the poster while a group of the audience tours the displays. It is usual for the author to be given the opportunity to introduce the work in a 2 to 5 minute oral presentation.

The relative informality of the poster situation should not relieve a scientist of the responsibility for clear communication and a professional attitude. Your knowledge of the subject, your candor in discussing the science with others, and your appearance and attitude are important to the presentation. It is an opportunity to improve your visibility at the meeting and to project your professional image.

This chapter covers the planning and presentation skills necessary to convey the information present in the poster in a concise, interesting manner within such time constraints. The chapter also considers the aspects of consciously managing your appearance and behavior to make a professional impact. Much of the issues raised will be as relevant to your everyday professional life as to the presentational situation under discussion.

13.1. PREPARING FOR THE POSTER SESSION PRESENTATION

The poster session presentation is much less onerous if you have had the good sense to distil your research findings to the points you consider to be the most important and have prepared a script. Most important of all is that you have practiced the talk, and if necessary have adjusted the script so it will fit your allotted time. To effectively introduce yourself and your research in a discernable fashion within the time constraints of a poster session, I would recommend the following scheme:

1. Introduce yourself. It is likely that not all of your audience will understand what you do or what your job title means, so spend the first few moments explaining your role in your organization and the relevance of your research. This is also an opportunity to promote yourself, so take a little time to ensure that you include some of your personal aims and career aspirations.
2. Introduce your research. Spend the next 30 seconds explaining just enough to justify your research and to present the aims of your study.
3. State your conclusions. Restrict your conclusions to those specifically pertinent to the aims of your study. Depending on the number of points you need to make, allow about 30 seconds to clearly make each point and about the same time to refer to the relevant section in your poster.
4. Give an overview of the findings and deliver the take-home message. The last part of your talk should be covered within about a minute, with emphasis being put on the take-home message.

13.2. PRESENTATION SKILLS

Many publications are available that consider in depth the attributes that contribute to skilled public speaking. I have listed some publications to which I would recommend you refer in the further reading section of this book. In this chapter I confine my deliberations to some of the major aspects of presenting a short overview in the environment of a poster session.

Many scientists dread the prospect of speaking in public but accept that the ability to communicate well at all levels is important in maintaining professional credibility. It is possible that you have chosen to present your research findings in a poster format deliberately to avoid a more formal spoken presentation. If this is the case I would urge you to use the poster session presentation as a training ground to present yourself well and with confidence.

You need to create a relationship with the audience. Remember you are among friends. Smile and try to appear at ease. The format of a poster session will be less formal than a traditional oral presentation, and your audience will probably be crowded around your display in very close proximity. It should therefore be easy to establish eye contact, one of the best ways of establishing contact with your audience. During your talk keep your head up, periodically scanning the audience and making eye contact for two or three seconds when possible. Although your presentation will be short, you should put some variety into your voice. Changing the pace, the volume, and inserting pauses into your delivery helps to maintain the audience's interest.

13.2.1. Rehearsing

Although the poster session oral presentation is short it should still be approached seriously. There is no substitute for rehearsing your spoken presentation. It should be practiced speaking aloud, while standing beside your poster, so you can familiarize yourself with the sections to be highlighted to your audience. Know your poster. Make sure you are able to locate any piece of information on the display quickly. You may find a pocket telescopic pointer or laser pen a useful adjunct. Ideally, record your rehearsal using a video recorder or, if this is not available, a tape recorder. Listen critically to your voice. Is it clear? Does it sound interesting? Most important, is there enthusiasm and vitality in your voice? If not, think of ways to improve the script or the delivery of the presentation and practice repeatedly. Rehearse in front of friends and colleagues and obtain feedback from them. It is only though such rehearsals and fine tuning of your script that you will approach the poster session with confidence and peace of mind. Rehearsing helps you to become familiar with your material, feel more confident, control nerves, develop an effective speaking voice, and use positive and appropriate body language.

13.3. PROJECTING A PROFESSIONAL IMAGE

Many people, scientists in particular, have an obstinate resistance to the idea of consciously creating an image for themselves. Intellectually, we would all like to think that we are not seduced by packaging or advertising, preferring to think that we are persuaded by rational, objective judgement of a product, service, or person. The fact of the matter is that research has shown that nearly all of us make decisions about each other within a few minutes of meeting. An individual's physical appearance, ethnicity, age, sex, height, weight, coloring, voice characteristics, body language, clothes, and grooming all make a contribution. So like it or not, we all create some sort of impression, even if it is the impression that we do not care about our image. If you do not bother to manage the impression you create you run the risk of being misinterpreted. People will judge you on the way you dress, the way you speak, the way you order your environment, and the way you interact with colleagues whether you wish them to or not.

People judge on first impressions. If you look significant, they assume that you are, unless you prove them wrong. If you look inconsequential, you will then have to struggle to overcome prejudices and barriers to prove that you are someone who should be taken seriously.

To create a lasting impression you need to stand out from the crowd, if only in some minor way. You need to leave your various audiences with an impression of you that they will remember, a trademark, and a distinctive look that will make you instantly recognizable. Using clothes as badges of office is as old as civilization. Start to cultivate your image by looking dispassionately at yourself. If you are not honest with yourself about your weak points as well as your strong ones, you will not be able to improve on them or correct them. You need to find a style that conveys the right messages about you, with which you feel confident, and that suits you.

13.3.1. Looking Like a Professional

Body language, clothes, and grooming are visual aspects of your professional image that are under your control.

13.3.1a. Body Language. However you look or move, you will be giving signs to your audience regarding your inner feelings. Once you are aware of your particular mannerisms and nervous gestures, you will be able to correct them to present a confident, unself-conscious image. Fidgety feet; fiddling fingers; clutching, patting, wringing hands; and not looking at the audience are just some of the telltale signs of nervousness that your body language may reveal. During your poster session presentation, consciously make an effort to do the following:

- make eye contact with your audience
- smile
- keep your hands empty and still
- start gestures from the shoulder and keep them relevant
- avoid hiding behind barriers, such as a table
- stand tall; do not sit
- balance on both feet, placed firmly on the ground
- appear calm and confident

13.3.1b. Dress-sense. Dress is one way of getting people to take you seriously. It is not my intention in this book to preach the benefits of adopting "correct" business dress, as the definition of "correct" will vary from company to company, from country to country, and of course between individuals. It should go without saying that for the entire meeting you should dress in an appropriate manner for the occasion as befits a professional. Clean, tidy clothes in which you feel comfortable, but those in which you would also be content to meet your superiors for a business lunch should be worn. It is important to consider the impact that your dress sense may have on your professional image, particularly during times of high visibility such as the poster round session.

There are two stages in choosing what you are going to wear when you are in front of your audiences. The first involves conforming to the basic uniform in which they expect to see you. The second involves making some dramatic alteration to your appearance that marks you as an individual and makes you memorable. When people first see you, they need to feel comfortable that you are what they expected—a successful and innovative scientist. As such I would recommend adopting a traditional "establishment" business uniform. This style consists of a dark suit in navy or gray worn with white or light linen shirt; small doses of color in the shirt, tie, blouse, or scarf; not much in the way of pattern and texture; dark shoes and dark hose; discreet jewelry; clean lines with a minimum of decoration; and classic rather than fashionable styling. Many readers may consider this too formal for scientific meetings where it is not unusual for a jacket and trousers option to be adopted. This is of course fine, providing you feel comfortable with your choice, and that you have found a style that suits your personality and professional image. Whatever your choice of style, remember that quality clothes project an image of success.

There are some dress styles that are definitely to be avoided at such occasions. These include jeans, particularly if torn or faded; t-shirts, particularly if they have inappropriate slogans; track suits; trainers; tight or revealing clothes; strong, brightly colored clothes, particularly in uncoordinated combinations; large pieces of jewelry; and leggings. An individualistic aspect to your chosen dress style may be added in a variety of ways, for example, by always wearing a bow tie, a flower in your buttonhole, or a distinctive waistcoat. Care should be taken, however, to ensure that such elements are in keeping with the rest of your dress style or they may appear rather theatrical.

13.3.1c. Grooming. Poor grooming is associated with being a failure. It conveys the impression of someone with low self-esteem. Let's face it, if you appear to have little regard for your own personal worth it is highly unlikely that others are going to conclude differently. Good grooming is a habit easily acquired if you are sufficiently motivated. The condition of your teeth, skin, hair, nails, and posture indicates how much you value yourself and contributes to create the image of someone to be valued. Some of the most common turn-offs include bad breath; a careless shave; chipped nail polish; dandruff; an overpowering fragrance; a poor complexion; a strange body odor; greasy or dishevelled hair; uncared-for shoes; creased, dirty, or ill-matching items of clothing; dirty fingernails; slovenly posture; and chewing gum. Looking good makes you feel good, and you will convey this to your audience who will then respond positively to your presentation.

13.3.2. Sounding Like a Professional

As part of your preparation, you will have prepared and practiced a short oral presentation for the poster round session. You should ensure that you are familiar with the content of this and feel comfortable with your presentation. The way you sound in delivering your scientific message is important, as it will affect the projection of your overall professional image. Consider what you say and how you say it. You need to be clearly heard and lucid, and although you are an expert in your particular field, you must be open to constructive criticism of your work. You must avoid at all costs being rude, dismissive, or condescending to those commenting on your work, as these traits of arrogance will have a negative effect on your projected image. Some of the most common gaffes include the use of street language; mumbling; speaking too quietly; not being knowledgeable in your field; the excessive use of jargon; the use of cliches; the use of platitudes; swearing; the use of inappropriate jokes; the use of funny voices; the use of sarcasm; and moaning or complaining.

13.3.3. Behaving Like a Professional

I have stressed the importance of projecting a professional image at the poster presentation, but as well as looking the part, you must also act the part if you are going to maximize your potential.

13.3.3a. Responsibilities. The professional meeting will offer many distractions for you as well as for your audience. It is your responsibility to be with your poster whenever you are scheduled to be, as some of your audience will have made an effort to be there to talk with you. You may also face the distraction of friends and acquaintances, but you should not neglect your poster audience for the social audience.

You should not be disappointed or despondent if there may be periods when no one approaches your poster. It is not unusual for this to occasionally occur, but you should still remain at your post for the complete duration of your scheduled period. One important attraction of the poster technique is that the audience is limited to only the truly interested. If your poster adheres to the criteria for a good poster and if a few people read most of it and talk with you, you have been successful. Whether the audience is one or a dozen, execute your professional role with clear communication, knowledge, and sincerity.

13.3.3b. Attitude. You need to project a positive, professional attitude toward your research and your audience. Convey your enthusiasm through your tone of voice, your willingness to discuss your findings, and by showing obvious pleasure at being given the opportunity to present your work to the audience. Relax and enjoy the experience; after all you are among fellow scientists who are interested in what you have to say.

14

A GUIDE TO HOSTING POSTER SESSIONS

Hosting poster sessions is largely a matter of proper planning and gathering and disseminating information. Those hosting poster sessions have responsibilities to the scientists presenting their work, to the overall conference organizers, and above all to the audience. Planning and organizing the poster sessions should be an integral part of the overall meeting strategy.

From the outset, careful consideration should be given to the suitability of any potential meeting venue to accommodate poster displays. This should be part of the criteria in selecting the most suitable meeting venue. The predicted number of delegates will generally provide an estimation of the number of posters likely to be offered for display and the number and size of rooms required. However, although this may act as a guide it is likely that the venues available from which to choose will ultimately dictate the number of posters that may be accepted for display.

Some questions you may wish to ask when considering the suitability of accommodation for poster displays include the following:

- Are there enough separate rooms available for the poster displays?
- Are the rooms in close vicinity to the main meeting rooms?
- Is there wheelchair access?
- Are the rooms large enough to accommodate both the poster displays and the audience and to allow the presentations to be made and viewed in comfort?

- Is there adequate lighting?
- Are the rooms accessible during the whole meeting?
- Are there tables and chairs available?
- Can coffee break refreshments be supplied or taken into the rooms?
- Are there toilet facilities nearby?
- Are sufficient display boards available?
- Are the display boards suitable, that is, of the same size and type, using the same fixings (pins, Velcro, etc.) for mounting the displays?
- Are there restrictions on the use of the rooms in terms of security or of health and safety?

Once you are content with the meeting venue and the accommodation for the poster displays you are in a position to commence planning the details of the poster sessions.

As the host of the poster session, you will need to have a clear understanding of the expectations of the overall meeting organizers. The number of posters to be accepted for display, the format of the poster abstracts, the scheduling of poster sessions, and so forth will need to be carefully planned, agreed on, and integrated into the overall meeting campaign. All of the information concerning the poster displays, however, should also be concurrently drawn together into a separate document. This will provide a clearer view of the proposed poster sessions and will enable vital checks to be more easily conducted.

Aspects of the poster sessions that need to be clearly defined include the number of posters, the format of the poster abstract, the grouping of posters into subject areas, the numbering system, and the dates and times of poster sessions.

Once you have a clear picture of the form in which the sessions are to be managed you will need to gather the more detailed information required by the poster presenters. This information should ideally be presented in an information sheet and should be supplied as part of the meeting's registration papers.

You may find the following checklist of the information that should be made available to the poster presenters in advance of the meeting useful.

Poster abstract:

- Notification of the space available for the abstract, or supply of a template
- Restrictions on the number of words

- Whether or not the abstract should be "camera ready" for publication
- Details of the preferred layout, typeface, and type size
- Where the abstracts are to be published
- Deadline for abstract submission

Poster displays:

- Dimensions, type, and color of display boards
- Restrictions on the dimensions of the poster
- Means of securing posters to the boards
- Availability of pins, Velcro, and so forth at the venue
- Number allocated to the poster
- Means of locating the building and room in which the poster is to be displayed
- Lighting conditions in the room
- Times that the poster room is accessible for mounting and removing poster displays
- Allocated poster session times when the authors should be present
- Schedule for poster display

Your role as host of the poster session does not end with finalizing the arrangements and receipt of the abstracts. It is most important that you maintain regular contact with the venue managers to ensure that the status quo is maintained. Any unforeseen changes that may effect poster presentations should be relayed to the contributors as soon as is practical. Additionally, whenever possible during the meeting, you should make yourself available to presenters to answer any queries or to resolve any specific problems.

15

COMMON POSTER DESIGN BLUNDERS

This chapter is included to encourage you to not only attend poster sessions and view as many displays as possible but also to develop a critical eye. In the future, after absorbing the scientific content of posters, try to reserve a minute or two to access the quality of the poster design. In the previous chapters we considered various aspects of poster design, production, and presentation. Bear these in mind as you view poster displays, and decide for yourself whether a particular style or combination of different design elements works. Is the scientific message clearly discernible? Is the display attractive and informative? Was it easily found in the poster session and did it have "pulling power"? Finally, above all, could you improve it or do better?

The following section illustrates some of the most common poster design gaffes. In doing so I have drawn on an extensive library of photographs that I have accumulated of posters displayed at many scientific meetings. To address the balance, however, I have also included some examples that exemplify the advice offered in earlier chapters. Before continuing I should make it clear that the posters chosen for illustration in this book are chosen because they are representative of many others. They were all presented at international scientific meetings. The comments I make are my personal opinion and they are made in the spirit of constructive criticism. With this in mind, and to save any personal embarrassment, some of the posters have been altered to protect the presenter's identity.

The illustrations themselves are of a quality that reflects the poorly illuminated, cramped conditions in which the displays are so often made. However, it is not my intention that you should necessarily be able to read the scientific content, but that it acts as an illustration of common design faults. Some of the most common design faults are considered in the following sections.

15.1. THE TITLE IS LARGER THAN NECESSARY

The posters featured in Figures 15.1 and 15.2 clearly illustrate this. Keeping titles concise and paying attention to the dimensions of the poster and of the display board supplied by the meeting organizers should avoid the problem.

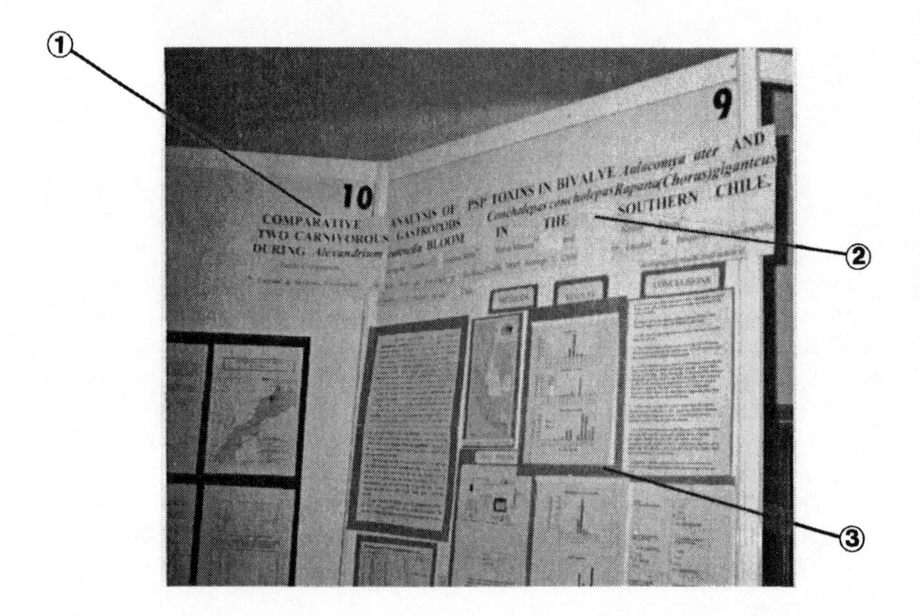

Figure 15.1. Poster critique (1)
Comments: The title of this poster is a good illustration of all of the things that could possibly be wrong! It is too large, overlapping both sides of the display board ①; large gaps appear as "rivers" between words ②; and it consists of a mix of words, some made up of capitals and others in lowercase letters. Even from just the top portion of the poster, it can be seen that the graphic and textual elements are crowded together ③.

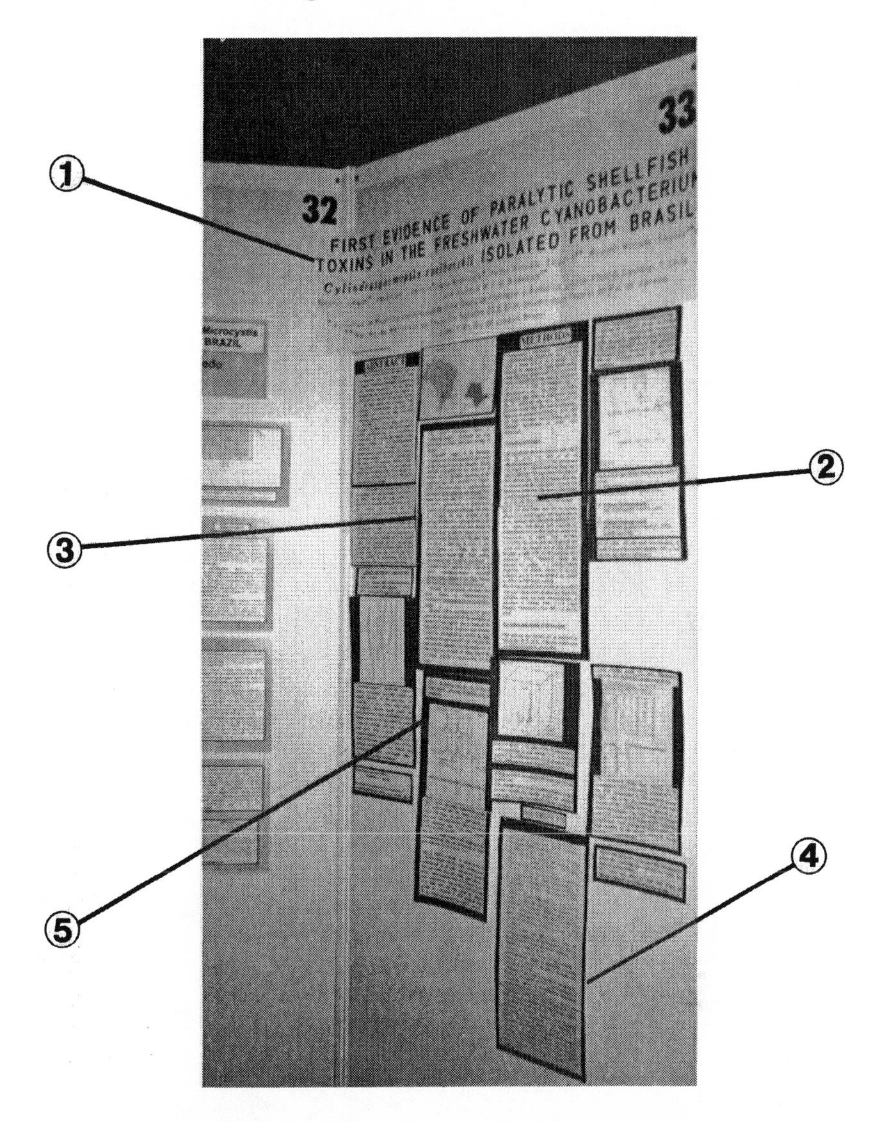

Figure 15.2. Poster critique (2)
Comments: This poster design has some serious faults. The title is too large for the display board ①; the textual elements consist of long, uninterrupted sections of text ②; the elements are crowded together ③; varying column sizes have produced a jagged appearance ④; and there is irregular spacing between elements ⑤.

15.2. THE AUTHOR'S NAMES AND AFFILIATIONS ARE TOO SMALL FOR READING FROM AFAR

This is a particularly common design fault that is illustrated in the majority of the posters shown in this chapter. A notable exception, with respect to the authors at least, is shown in Figure 15.3.

15.3. THE TEXT IS OVERLY WORDY

Too many poster presenters seem unable to grasp the idea that posters are a predominantly visual means of communicating. Large segments of un-

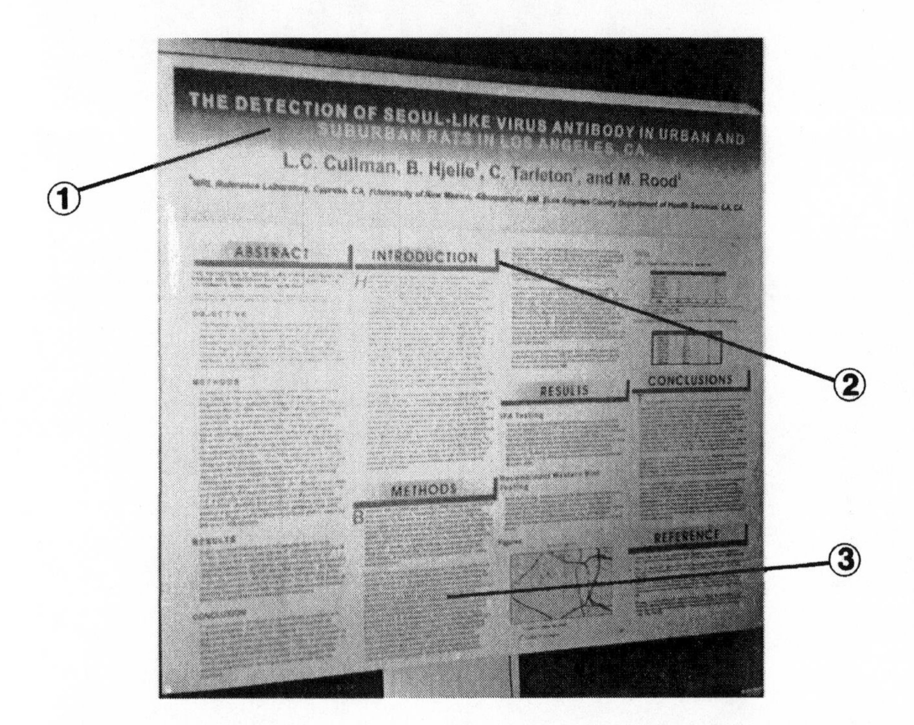

Figure 15.3. Poster critique (3)
Comments: This poster is an example of a well-organized, unified display, where unfortunately, the scientific content is visually "sterile." Although the title, authors ① and headings ② can be clearly seen, and the sections are in a logical order, the display comprises of long, uninterrupted sections of text ③! This type of poster is more often than not walked past by all but those specifically seeking the information.

broken text on posters (See figure 15.4) are not visually exciting, and they take a considerable effort to digest.

15.4. TEXT TYPE IS TOO SMALL OR TOO LARGE

Long lines of small type have a gray, "I am going to have to stand here for some considerable time to get through this!" appearance (see Figure 5.5).

I am going to have to stand here for some considerable time to get through this. I am going to have to stand here for some considerable time to get through this. I am going to have to stand here for some considerable time to get through this. I am going to have to stand here for some considerable time to get through this.

At the other end of the scale, short lines of large type also create hard-to-read textual elements. Excessive hyphenation is frequently used to avoid large gaps when type is too large for line length, particularly when using a justified text style.

Excessive hyphenat-
ion frequently res-
ults when presenti-
ng short lines of
large, fully justif-
ied type.

To make matters worse they are also extremely difficult to read. (Advice on the target number of words/characters per line is given in chapter 6.)

15.5. LACK OF UNITY OF STYLE

It is important to maintain a sense of unity of style within the poster so the individual elements can blend and merge to form the overall design. This aspect is frequently forgotten by poster presenters (see Figure 15.6).

15.6. DIRECTIONAL FLOW OF INFORMATION IS NOT CLEAR

To convey your message, the reading sequence of the elements in your poster needs to be easily discernible. Too often navigating a poster is made

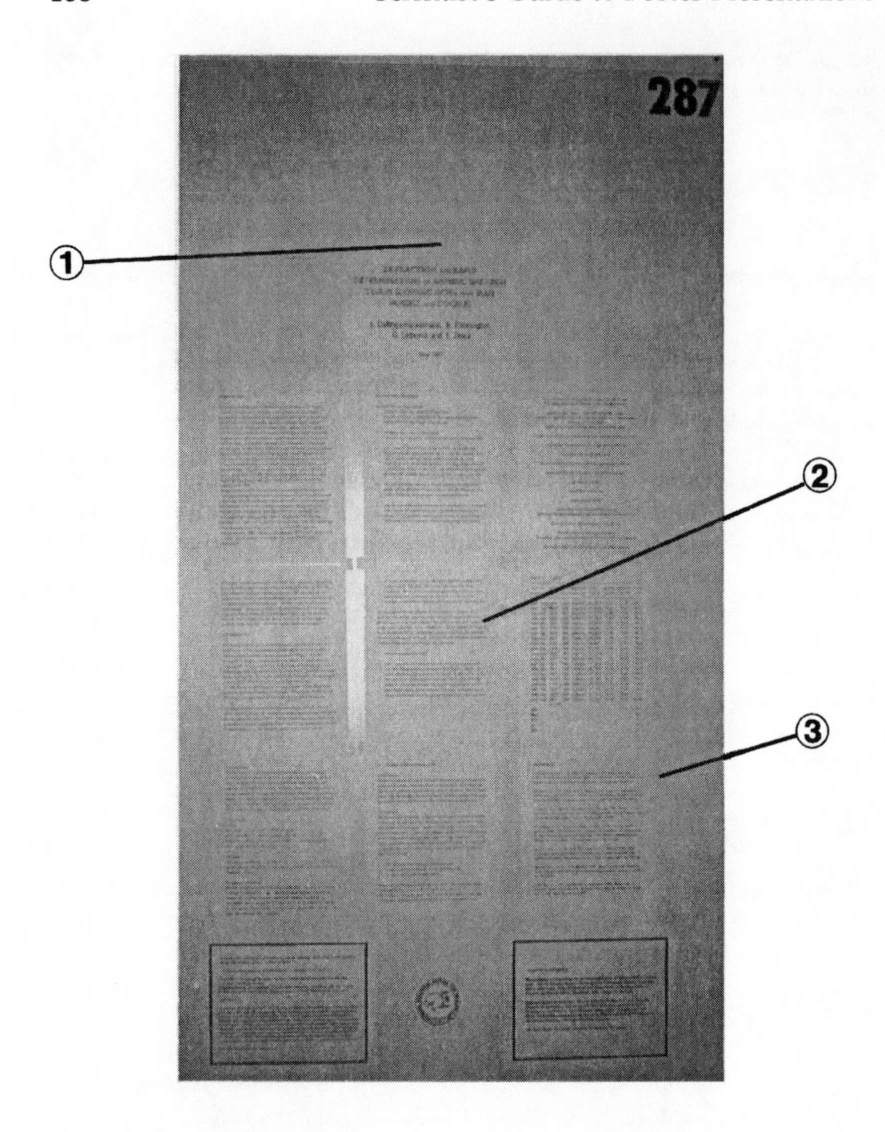

Figure 15.4. Poster critique (4)
Comments: This display is positioned too low on the poster board ①. The overall appearance is similar to Figure 15.3 in that it too is visually sterile. The display does not contain a single graphic element or illustration to break up the long, diffuse sections of uninterrupted text ②. To make matters worse, the text is printed on to white paper and has been mounted without use of borders or backing material onto a white display board, giving a diffuse grey hue ③.

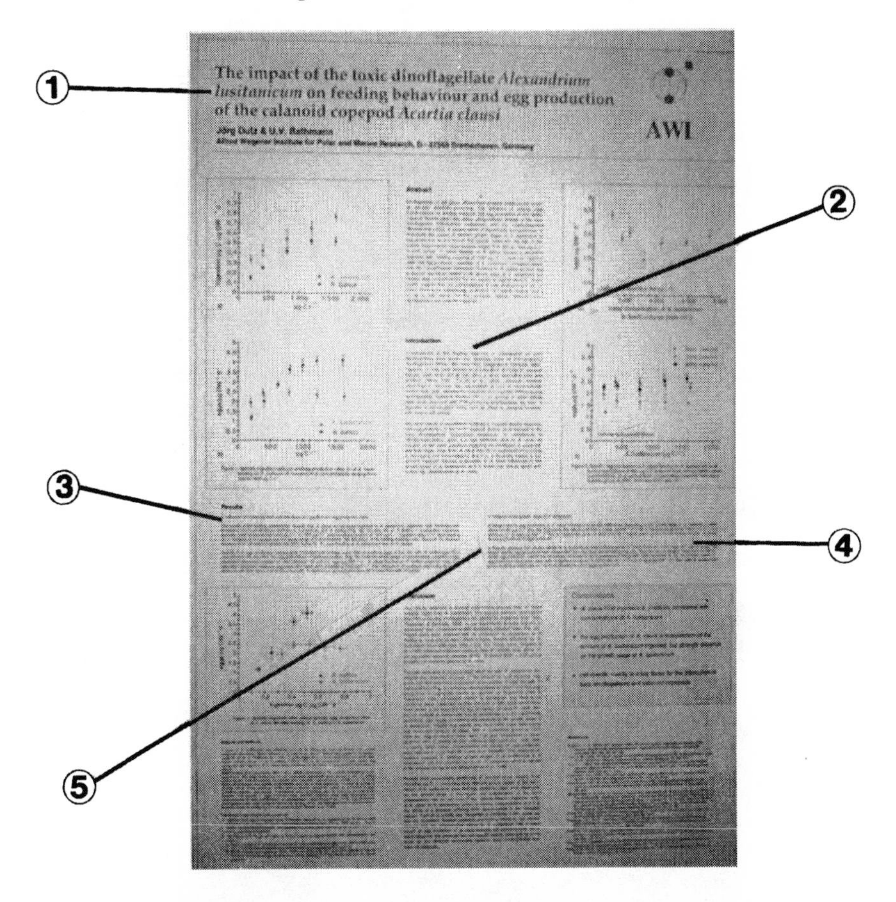

Figure 15.5. Poster critique (5)
Comments: The title of this display is clear ① but the layout of the design presents viewers with some difficulty in reading. The introduction is positioned in the central column ②, but subsequent textual elements are found in all three columns ③. Sections of long lines of text are included ④. To add to the confusion, column numbers change in the middle of the poster ⑤.

difficult as a result of section headings being too small or too large relative to the text type or absent altogether. Graphic elements that transverse the whole of the poster, effectively splitting it in two, often confuse the reading sequence (see Figure 15.7). Difficulties in following the chain of information also arise when the layout creates a reading sequence that criss-crosses the poster.

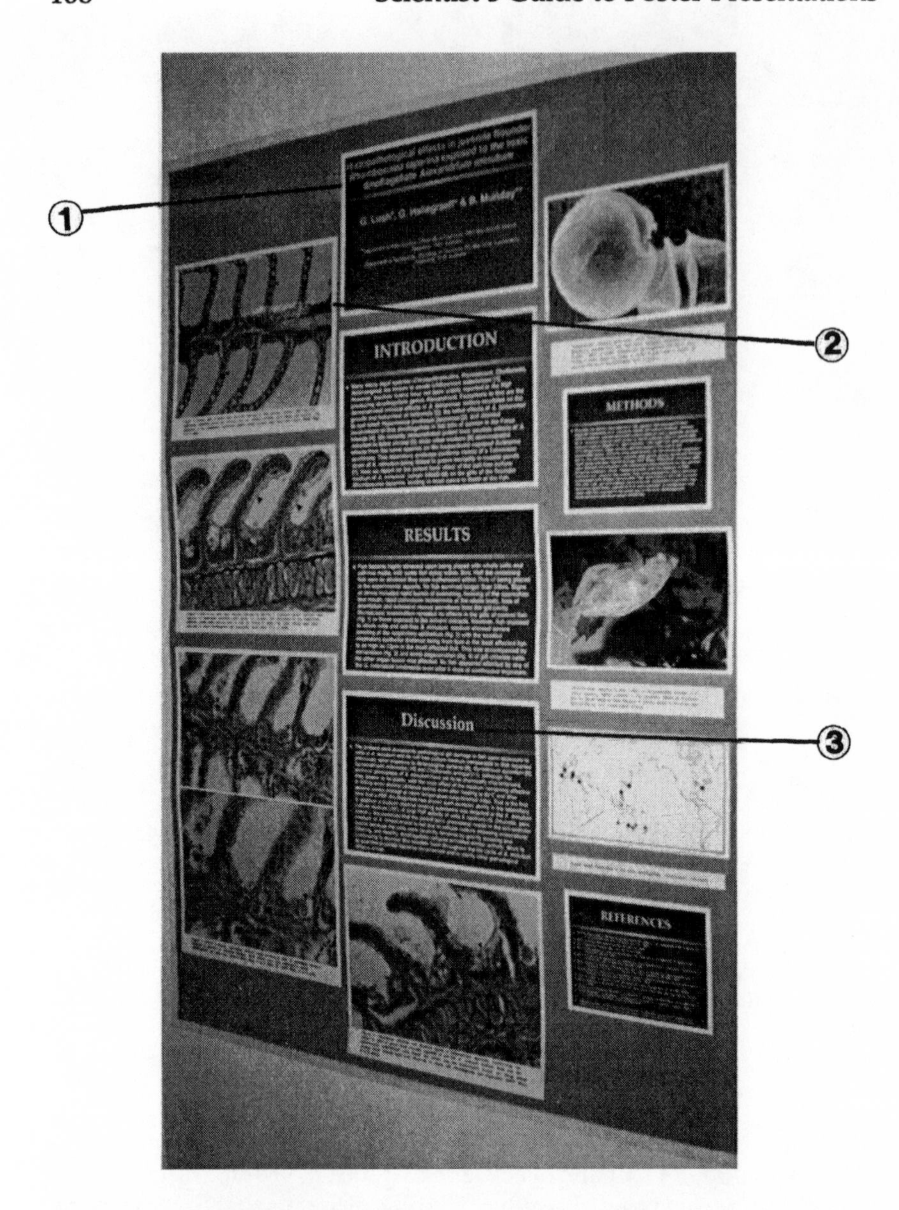

Figure 15.6. Poster critique (6)
Comments: The title of this poster is unlikely to attract attention ①. It is far too small! There are also some problems with the layout. As the first column consists of graphic elements ②, the viewer's eye has to jump from the central textual elements to the left and to the right to follow the flow of information. Additionally, there is a lack of unity in the use of capitals and lowercase letters in the headings ③.

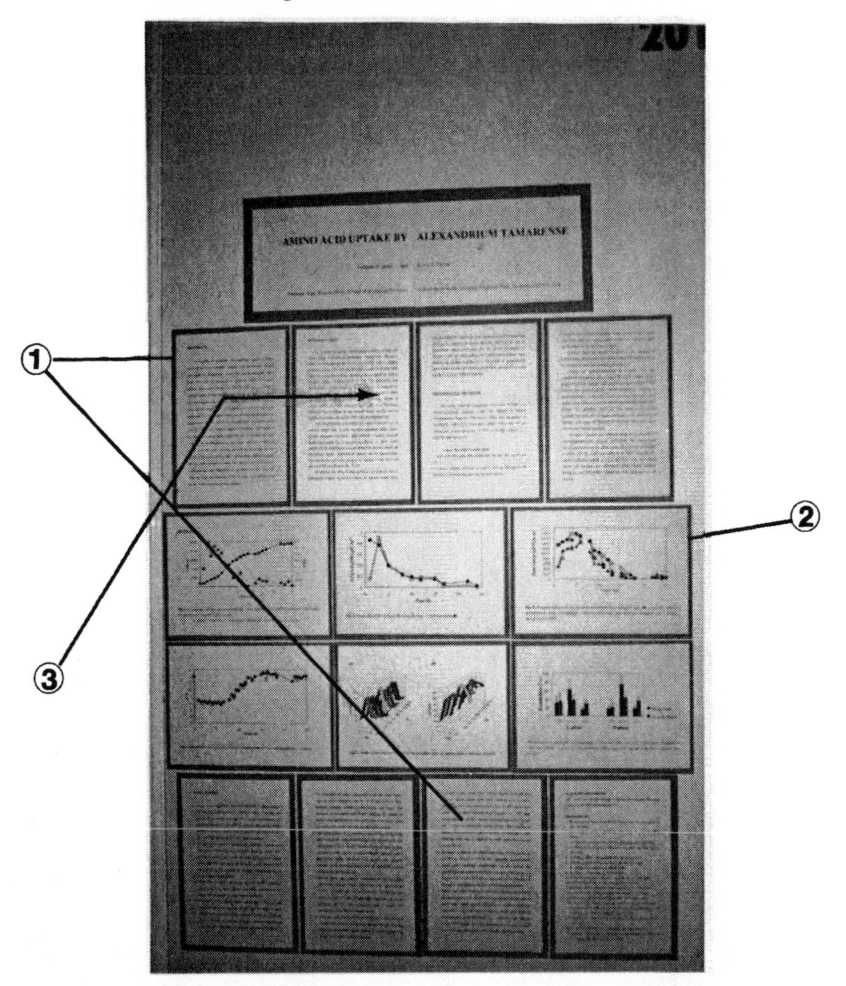

Figure 15.7. Poster critique (7)
Comments: The unified appearance of this poster suffers from the fact that it is arranged in such a manner that it consists of two rows of text ① totally separated by a row of illustrations ②. The flow of information is further complicated as the text reads from left to right across the columns ③.

15.7. POOR COMPOSITION

Posters are often presented that have no consistency of spacing or of placement of analogous elements, of which there are frequently too many (see Figure 15.8). The composition of such posters often takes on a jagged appearance with few long lines to carry the eye.

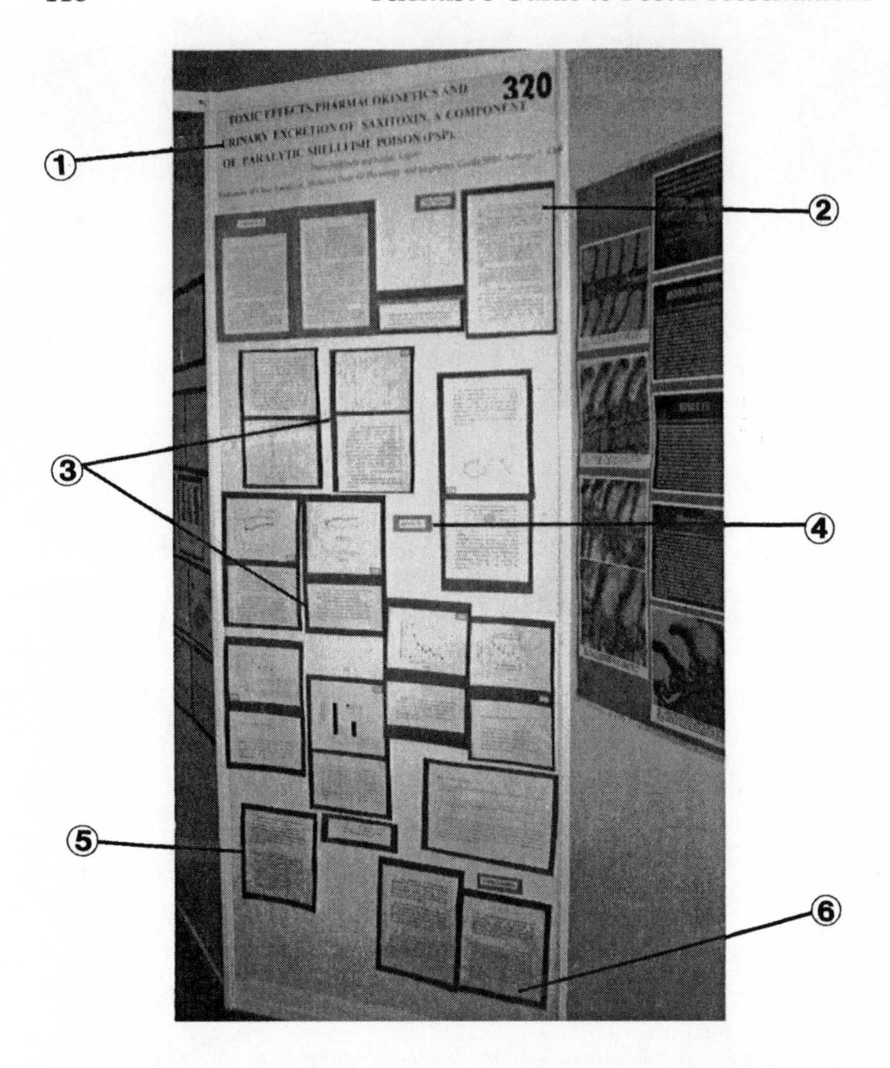

Figure 15.8. Poster critique (8)
Comments: This display has a muddled, messy appearance. The title, although rightly placed high on the display board, is too long ①, consisting of three lines of text. Many elements have been placed too high ②, or too low ⑤, ⑥ to read. Inconsistent use of blank space between elements ③ has resulted in confused column layout. Additionally, headings appear to float in space ④ rather than being associated with any particular section.

15.8. USING TOO MANY SEPARATE SMALL ELEMENTS, OR DIVIDING ELEMENTS INTO TOO MANY SMALL PARTS

Posters that contain too many individual elements tend to have a muddled appearance (see Figure 15.8). Similarly, larger elements that are divided into too many subunits may also appear confusing to the eye and and difficult to follow (see Figure 15.9). The latter is often seen in displays in which the authors are attempting to illustrate a trend. I would suggest that, as discussed in chapter 7, there are more effective means to illustrate such data at the poster designer's disposal.

15.9. INADEQUATE USE OF BLANK SPACE

Many poster presenters cannot resist the temptation to fill all available space (see Figure 15.7). As discussed in chapter 8, blank space is most important as it dictates the openness of the overall design and allows the eye to momentarily rest between the various textual and graphic elements. Figure 15.10 illustrates better use of space than many of the examples of designs featured. However, even this poster design could make better use of blank space between elements in the horizontal plane. Figure 15.11 is a further example of better use of space but also illustrates the dangers of placing too much space between headings or captions and the associated element. The various elements seem to "float" in an unconnected manner. Additionally, putting blank space around an element has the effect of making that section stand out from the rest of the poster. This can be seen in Figure 15.12, where, whether intentional or not, the "Objectives" textual element (top left) commands the viewer's initial attention.

15.10. TITLES THAT ARE TOO WORDY AND LINE SPACING THAT IS TOO LARGE

The title is a very important, but frequently neglected aspect of poster design. The importance of keeping the title concise, snappy, and relevant was discussed in chapter 8. Titles are necessarily composed of letters of a larger type size than the rest of the poster, and all too frequently consist solely of capitals. Where titles consist of more than one line of text it is not uncommon for poster designers to assign a "default" line space, which is often too large (see Figure 15.8). Although you may be able to use the

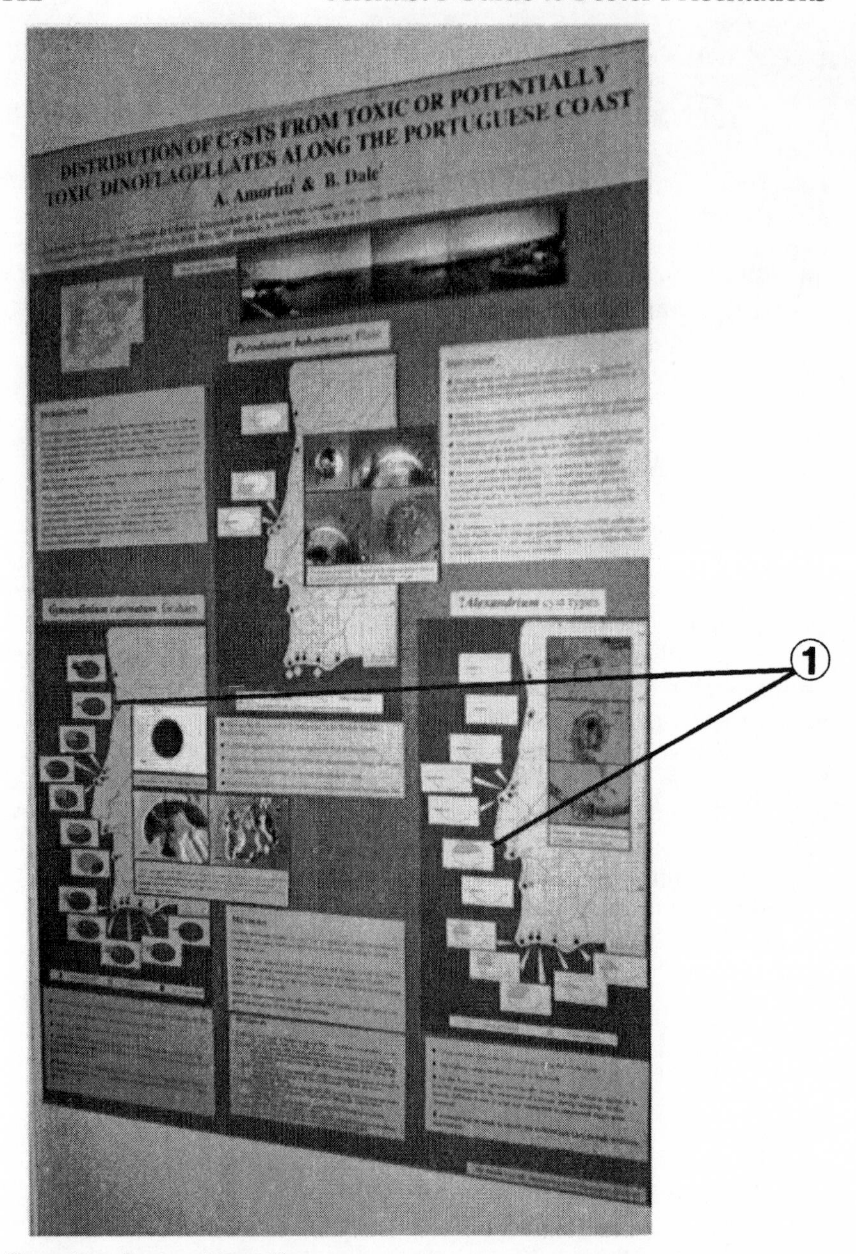

Figure 15.9. Poster critique (9)
Comments: This display illustrates the distracting, cluttered appearance created by using too many subdivisions within graphic elements ①.

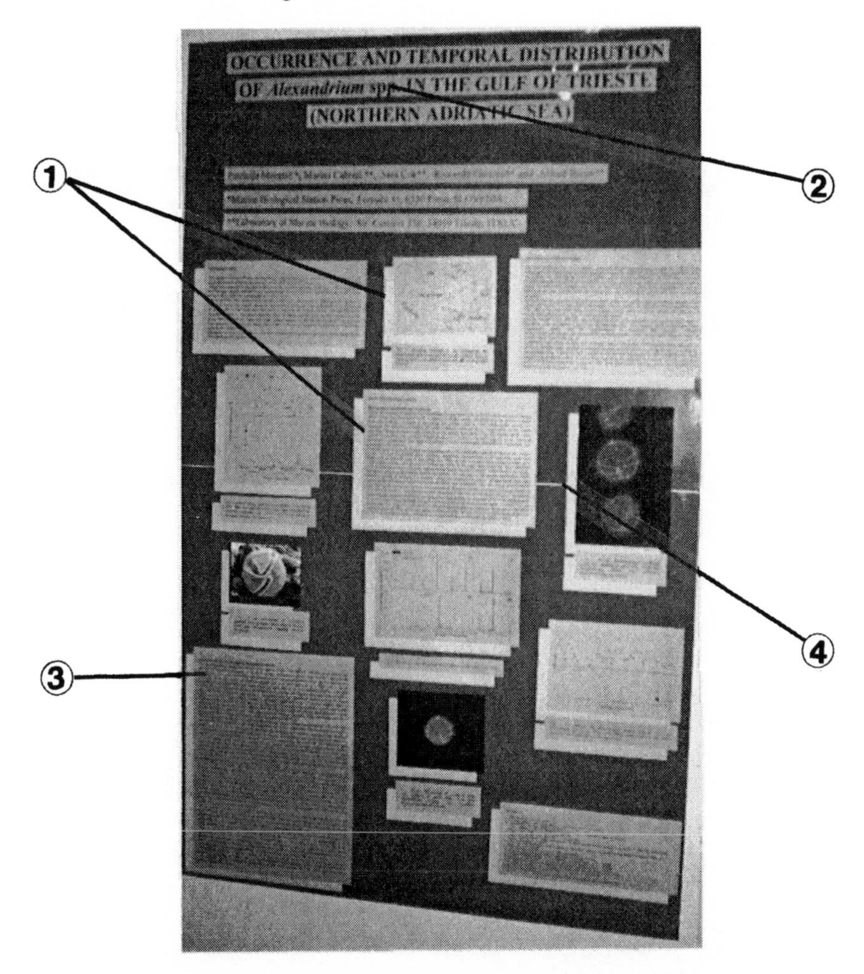

Figure 15.10. Poster critique (10)
Comments: This display makes effective use of shaded card to provide an interesting shadow effect to the textual and graphic elements ①. However, the title appears to lack unity, with capital letters for some words and lowercase for others ②, and textual elements consist of long sections of uninterrupted text ③. Additionally, the gap left between the two halves of the display upon mounting is distracting and appears to link elements in the second rows of the columns ④.

default line spacing in the main text of the poster, if used in titles the size of the type will result in lines set unnaturally far apart. For example:

"Default line spacing

with 12-point type"

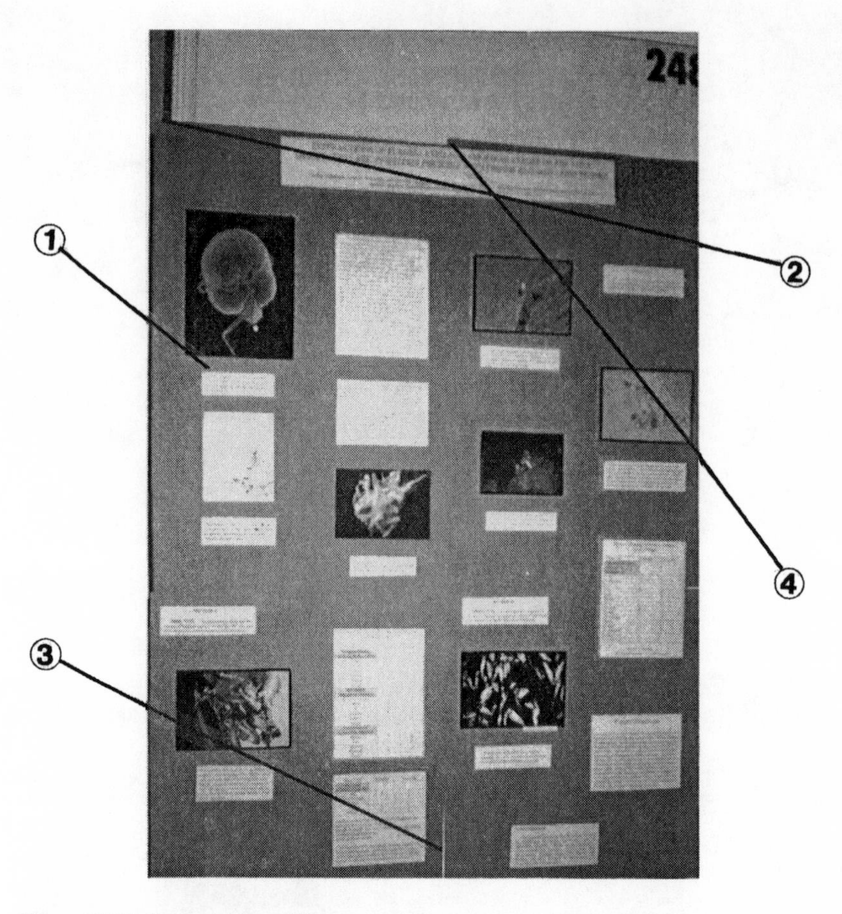

Figure 15.11. Poster critique (11)
Comments: The presenters of this display have attempted to make good use of blank space. Unfortunately, they have left too much space between captions and the elements with which they are associated ①. The result is a poster in which the various elements seem to float in an unconnected manner. There are also some problems with the mounting of this display. The poster overhangs the display board ②, and the sections are not properly aligned ③, resulting in distracting gaps and jagged edges ④.

"Default line spacing

with 36-point type"

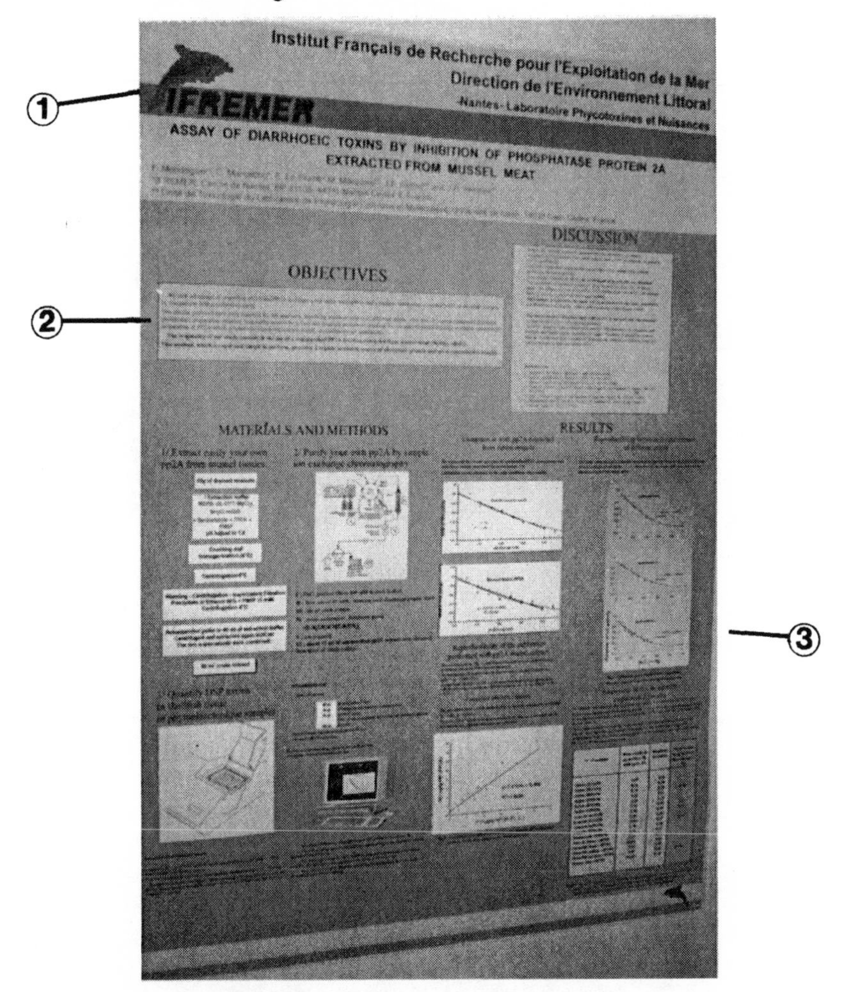

Figure 15.12. Poster critique (12)
Comments: This poster illustrates the effective use of a logo in the promotion of the author's affiliation ①. The display clearly highlights the top sections by placing more space around these elements ②, but this has left the lower elements rather crowded ③.

15.11. OVERPOWERING USE OF BACKGROUND GRAPHICS

Use of designs printed on the backing material of posters needs to be handled carefully. Figure 15.13 illustrates what happens when the design is too pronounced. It overpowers the textual and graphic elements of the

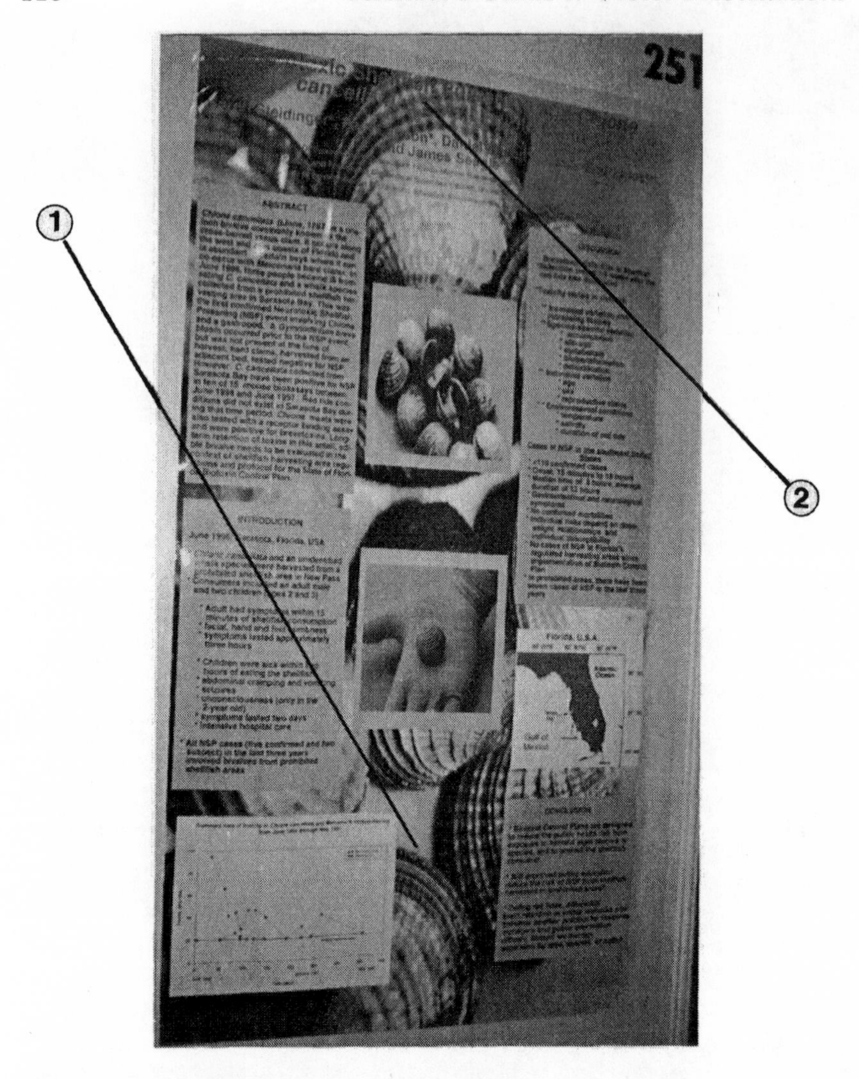

Figure 15.13. Poster critique (13)
Comments: This poster illustrates the effects of using an overpowering background design ①. It competes with the poster elements in a most distracting manner. Indeed it has made the title ② practically unreadable!

poster in a most distracting manner. Indeed, it almost totally obscures the title. A much more subtle approach, and one that works quite well, is shown in Figure 15.14.

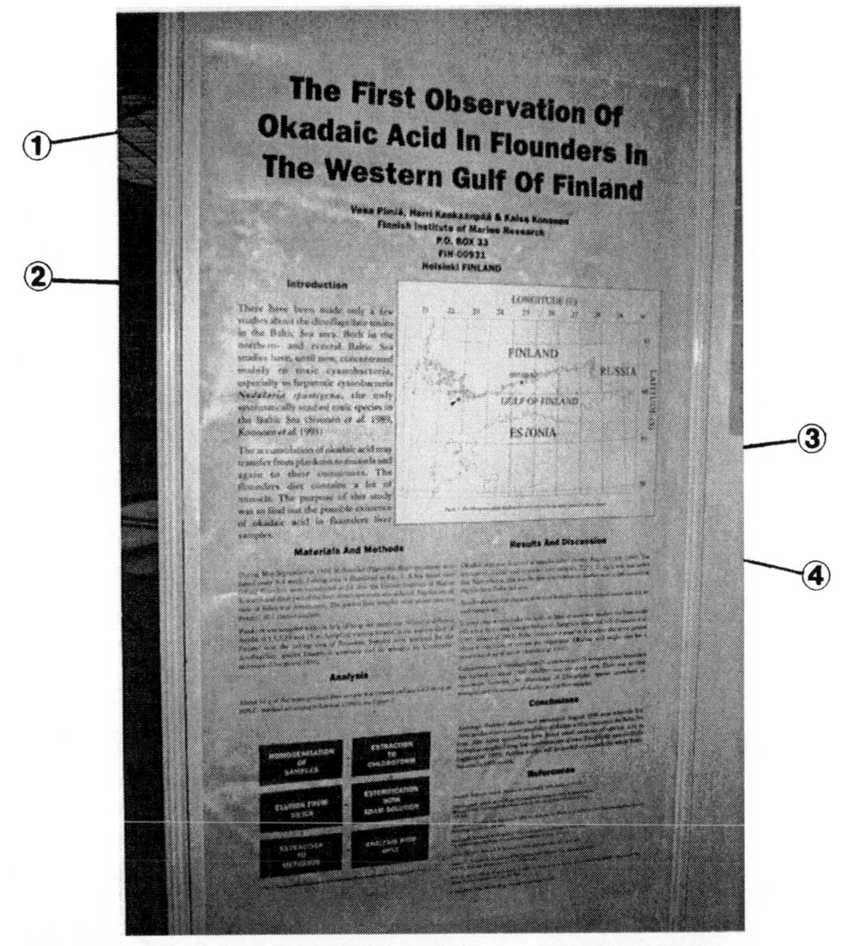

Figure 15.14. Poster critique (14)
Comments: This display has many attributes. The title is clear and easy to read ①. It is well organized, with the various sections clearly signposted ②. The choice of type size is particularly pleasing and easy to read ③. The tasteful use of gray shades and the inclusion of a discreet relevant background design ④ combine to unify this visually interesting poster.

15.12. MOUNTING POSTER ELEMENTS IN POSITIONS THAT MAKE THEM UNREADABLE

The desire to present as much information as possible often leads to this particularly common mistake (see Figure 15.15). Poster elements mounted below knee level or above head level are practically unreadable to the vast majority of viewers.

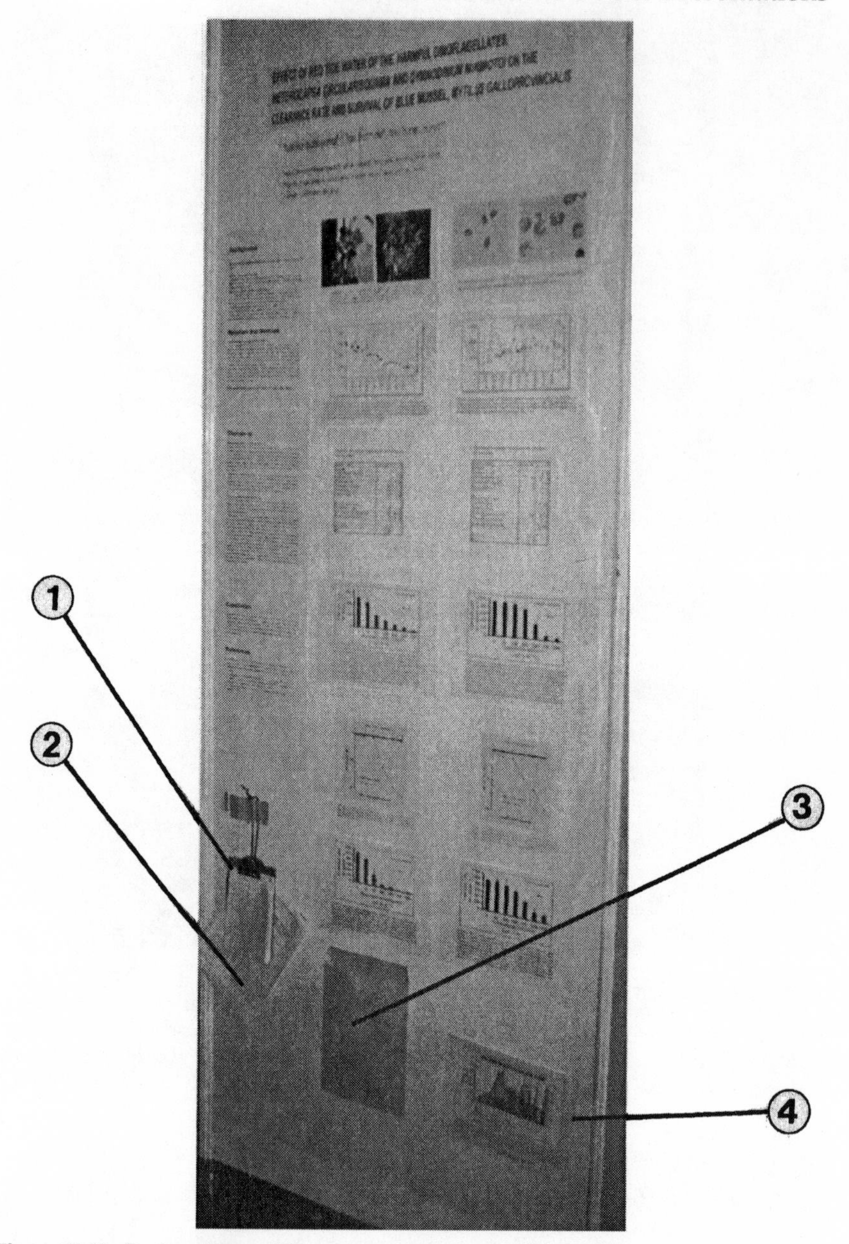

Figure 15.15. Poster critique (15)
Comments: The presenters of this poster are clearly aware of the potential value in communicating with their audience. They have usefully included a notepad ①, copies of scientific papers ②, and a contact envelope ③. Unfortunately, they have tried to display too many elements. The position of the lower ones makes them practically unreadable ④.

15.13. OVERZEALOUS USE OF COLOR

Color can make or break a poster display. The various ways in which color can be used was discussed in chapter 9. Figures 15.14 and 15.16 through 15.18 provide various examples that exemplify some of the advice offered in the earlier chapter.

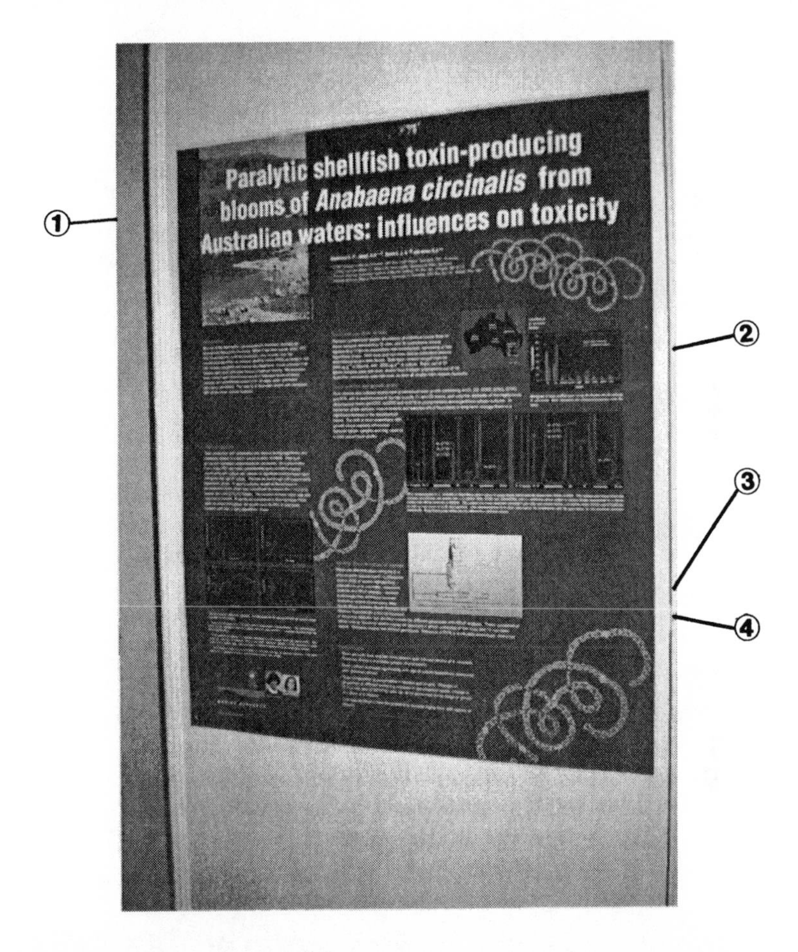

Figure 15.16. Poster critique (16)
Comments: What a beauty! This poster has very many attributes that have been brought together to produce a highly effective, professional-looking display. I will point out a few of these, but will leave you to recognize others that exemplify the comments made in earlier chapters. Among those attributes of particular note, I would include that the title stands out and can be clearly read ①; the impressive use of shading ②, and of background design ③; and the good use that has been made of blank space ④.

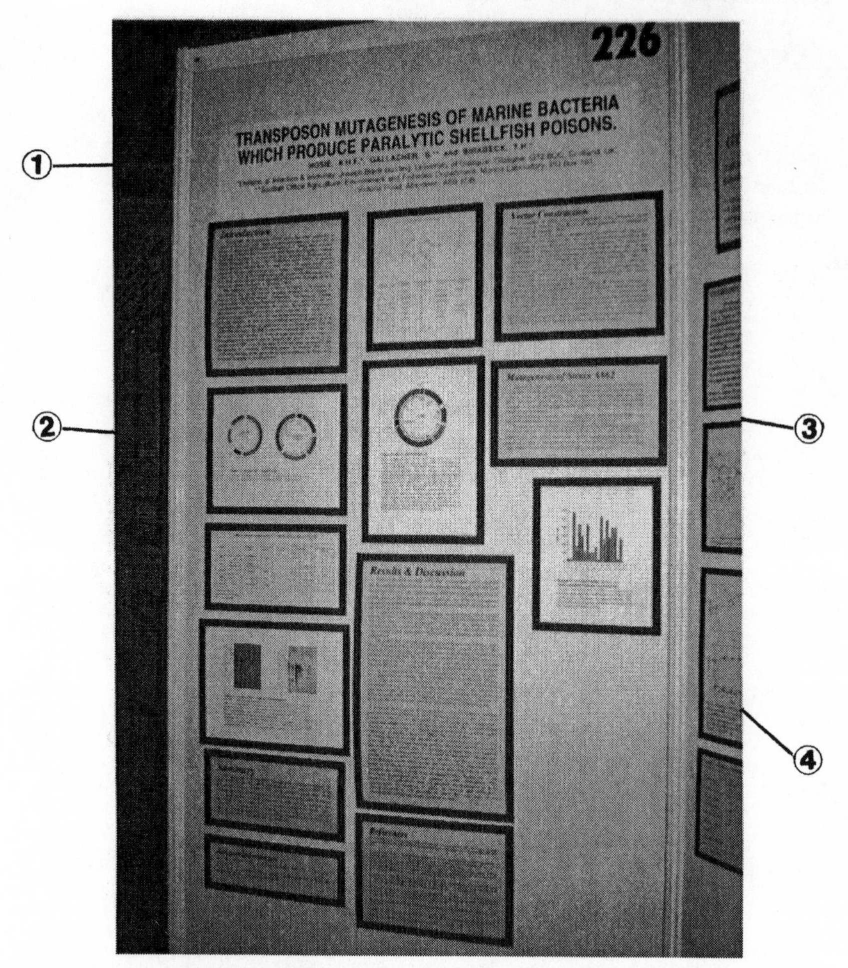

Figure 15.17. Poster critique (17)
Comments: The title of this poster is positioned high on the display board and can be clearly seen ①. The design makes good use of shading in graphic elements ②. Shading is also used as a background to distinguish textual elements ③, unity being preserved by the use of the same darker shaded border for all sections ④.

15.14. USE OF UNNECESSARY MATERIAL

Material that is not required to convey your scientific message or that does not add subtlety to the overall design should not be included in the poster design. Such material acts as a distraction. Figure 15.19 shows a poster

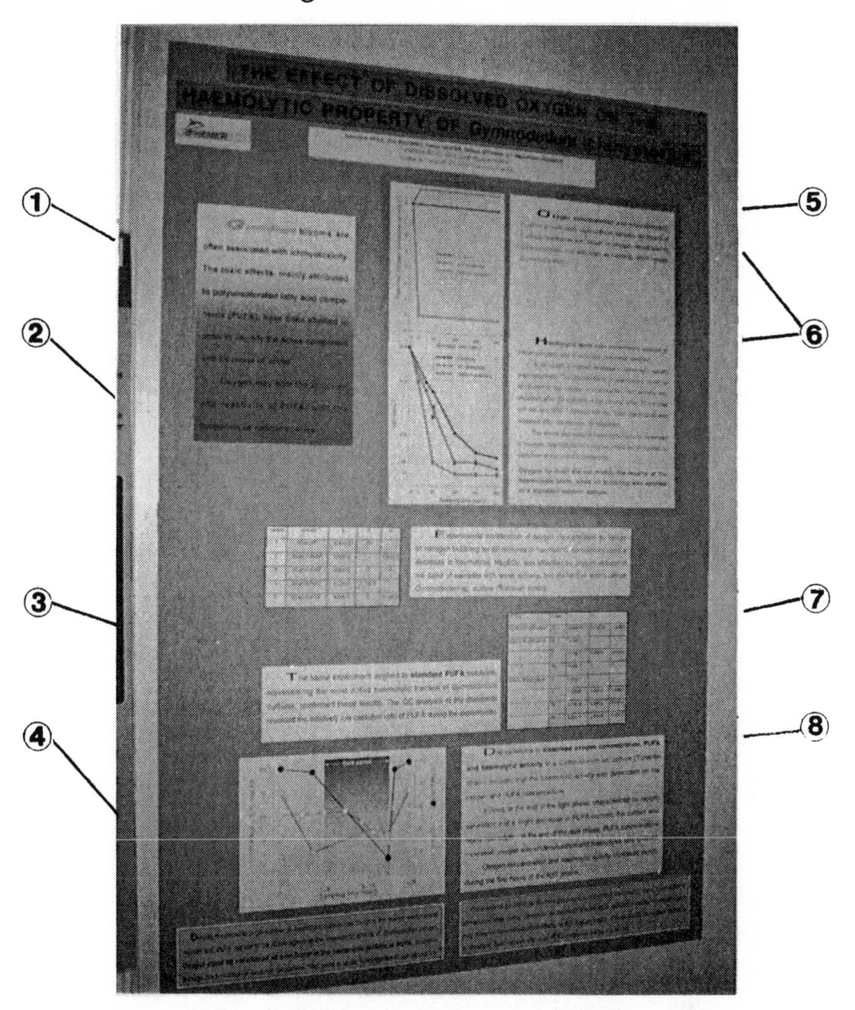

Figure 15.18. Poster critique (18)

Comments: This display shows several different ways in which shading may usefully be used in poster designs. Shading has been used to demarcate lines in graphs ①, highlight areas in tables ②, provide a background hue to areas of blank space ③, and to highlight particular textual elements ④. The chosen size of text enables the content to be easily read and raised capitals have been used to add interest at the start of sections ⑤. Unfortunately, however, the design lacks clear signs to guide the viewer. The title is masked by the use of too strong a background shade ⑥, the tables and graphs lack captions ⑦, and the organization of columns appears messy ⑧.

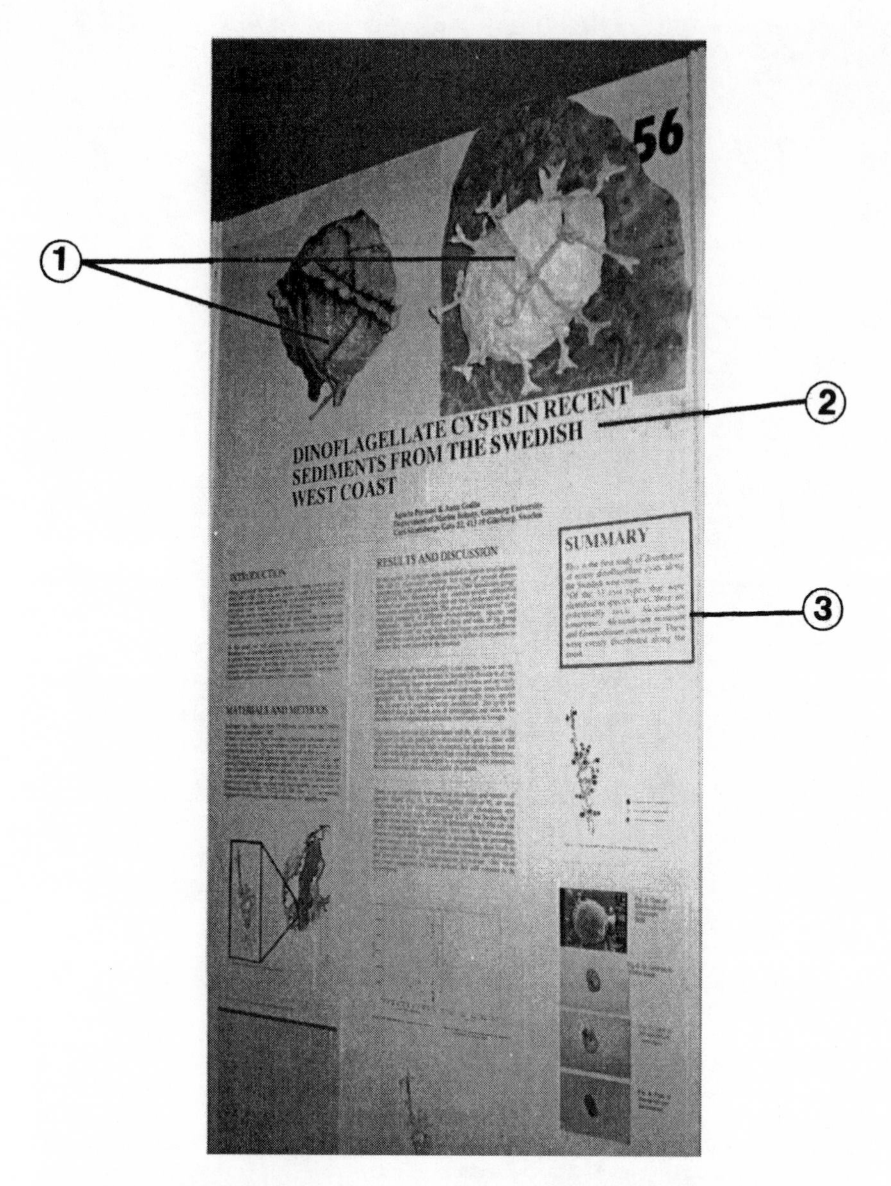

Figure 15.19. Poster critique (19)
Comments: This display makes use of three-dimensional models to attract attention ①. However, in doing so the title is positioned too low to be clearly seen from afar ②. The value of using a border to highlight a section is illustrated ③.

display that included three-dimensional models. Although effectively attracting attention to the poster, the models distract from the poster content. Additionally, their use may cause some viewers to doubt the professional credibility of the whole display.

15.15. POSTERS THAT ARE TOO BIG, TOO SMALL, OR THE WRONG SHAPE FOR THE DISPLAY BOARD

This basic error continues to occur to a certain degree at most poster sessions (see Figure 15.20). It is largely a result of presenters not paying attention to the meeting organizer's brief, but may also occur when the poster is being used at more than one scientific meeting.

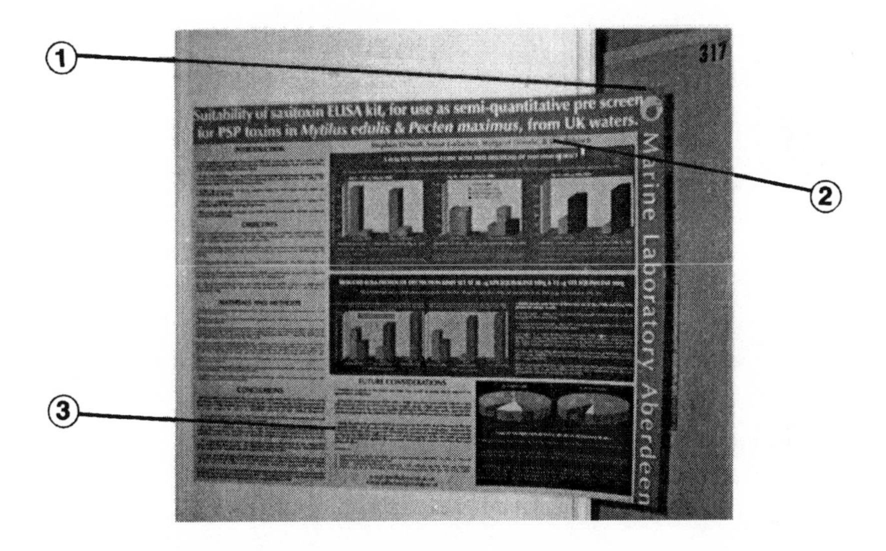

Figure 15.20. Poster critique (20)
Comments: This poster is the wrong shape for the display board and does not fit ①. The title is clear, but is positioned too close to the content of the poster ②. Textual elements consist of blocks of uninterrupted long lines of text ③.

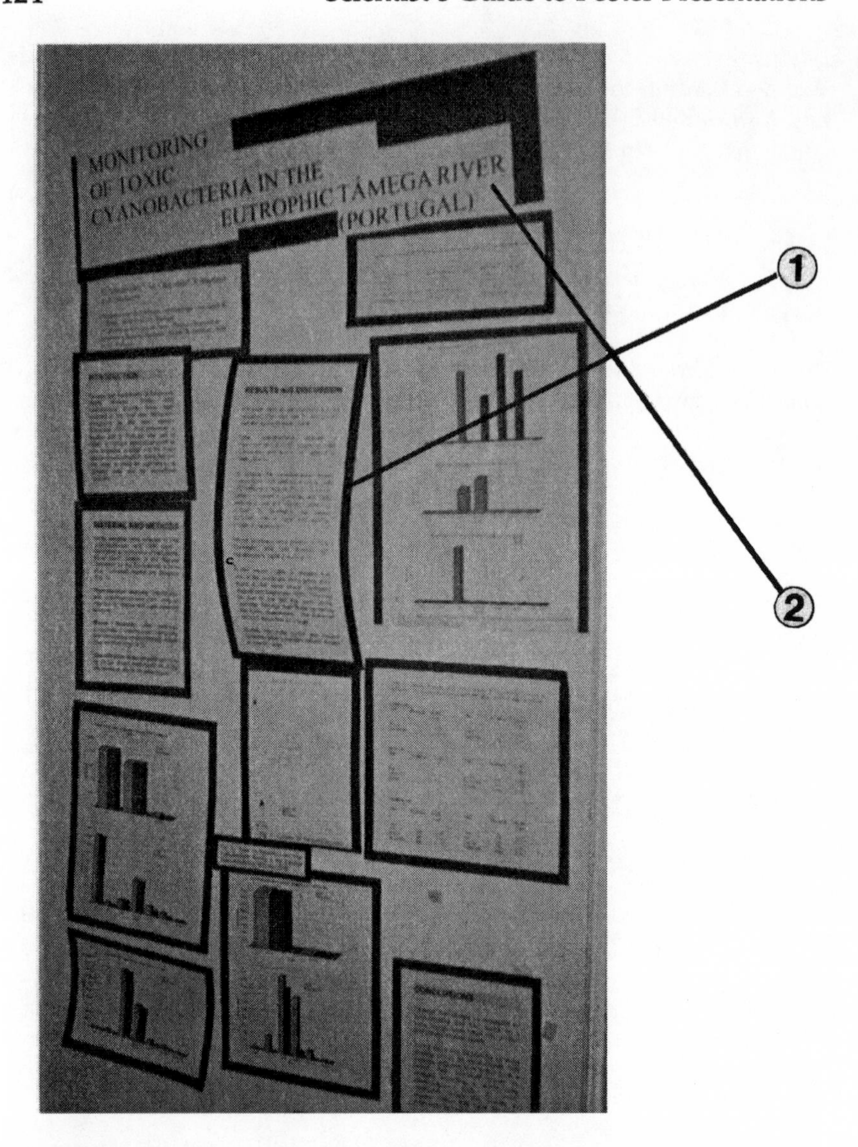

Figure 15.21. Poster critique (21)
Comments: This display has clearly been neglected ①, and has resulted in a dishevelled display that will almost certainly be ignored by viewers. The title of this poster appears almost as an afterthought ②. The sections making it up have been roughly "thrown" together. The words are complete but no thought has been given to aspects of design.

15.16. LACK OF ATTENTION ONCE INITIALLY MOUNTED

It should be stressed that a poster presenter's responsibility for their poster does not come to an end as soon as it is initially mounted. The condition of the poster should be periodically checked and any necessary adjustments or repairs made, if, for example, the sorry sight shown in Figure 15.21 is to be avoided. Many of these gaffes significantly reduce legibility, comprehension, or simply render the poster's message less accessible.

I hope that this critique of the posters illustrated will help you in identifying at an early stage potential pitfalls in your own poster design and presentation. Having identified any problems, you will be able to take corrective action before putting it on display.

I would like to conclude this chapter by emphasizing that the worst poster of all is the "no show" poster. Never offer a poster, or provide an abstract, if you are going to be unable to produce a poster to the required specifications in the time scale allocated.

16

FUTURE TRENDS

Predicting the future is notoriously unreliable. However, it seems reasonable to extrapolate current trends to consider likely developments in the foreseeable future. The 1980s and the 1990s have seen the emergence of scientific posters and have seen their popularity to grow to such an extent that they now represent the most frequently used means of communicating scientific information at many meetings. Over this time, the standards of poster design and quality of production have improved in leaps and bounds. This is partly due to the developments in, and availability of, desktop computers and the software supporting them, which has developed over the same period. The gap between the quality of amateur-produced posters and those professionally prepared is closing, but there is still much room for improvement. For many, the gap still remains a chasm. The desktop computer is already a most valuable asset in poster production, allowing text and graphic representation of scientific data to be readily prepared in color. Photographs and other illustrations may be scanned and incorporated directly into the poster design. It is probably only a matter of time before a computer software program becomes available specifically to cover all aspects of poster design and production. If this becomes available, poster modules could easily be transported via a floppy disc, and printed on arriving at the meeting venue, and subsequently constructed on the display board.

Future trends in visual presentation of scientific information in general are likely to change dramatically from 1999 to 2009. The cost and availability of video recording and playback equipment has now made it possible to easily communicate scientific information visually by video. Such presentations, if adapted to posters displays, could be projected directly from a portable computer, and this technology would have the added benefit that it allows spoken text to be added to the presentation. Computer-assisted slide presentations are already not infrequent in oral presentations. However, these forms of presentation depend on access to suitable equipment at the meeting venue and would rely on the use of headphones to avoid an almighty bombardment of sound in the meeting room. Perhaps most important, these means of conveying information lose the personal contact that is one of the major strengths of poster presentations. I do not predict that such technological advances will distract from the appeal of the traditional poster format.

The Internet, however, is a powerful form of communication that is impossible to ignore. It is capable of communicating with many thousands of interested individuals. I would predict that it will play a major role in future visual communications of scientific information. Many published journals are now accessible in their entirety, complete with visual graphics, on the Internet. Organizers of some large scientific meetings have posted the timetable of events, and have included complete texts of abstracts. I am sure that it will not be long before abstracts accessible on the Internet take on a much more detailed and elaborate format, possibly including a downloadable image of the whole poster. A note of the author's e-mail address would enable contact to be made. This would ensure that even those scientists who were unable to attend the meeting could be informed. It is common for viewers who are particularly interested in the content, format, or a graphic element of a poster to photograph either the whole thing or specific segments. It is already commonplace for presenters to supply delegates with color facsimiles of the poster, and this practice will almost certainly increase.

Recent attention has been drawn in the scientific press to the lack of peer review of the information offered through poster sessions. However, I can not foresee a rapid resolution to this, as by the very nature of the summarized information supplied on a poster it would be an impossible task to conduct such a review of all contributions submitted. In the same way as the quality of the information gathered from the Internet, so too must the information gleaned from posters be judged. This judgment may be based on factors such as the credibility of the author the credibility of their working establishment, and the feasibility of the scientific message together with how well the work appears to relate to what is already firmly established.

The future of the poster in its present format is ensured for the foreseeable future, and as the number of scientific meetings is increasing I am sure the value and popularity of the poster session will continue to grow.

GLOSSARY OF TERMS

A

Abstract: a concise summary of the basic content of the paper that can stand alone without presenting extensive experimental details.
Acronyms: words made from the initial letters of the full names, e.g. AIDS.
Affiliation: organization with which a person is associated.
Alignment: (right/left/center) lining up type or illustrations on a horizontal or vertical grid.
Ambiguity: lack of clearness through double meaning.

B

Backing material: the substance supporting the various poster elements.
Blank space: the background area surrounding the various poster elements.
Bleed: to diffuse into the paper and make feathery edges rather than crisp edges along lines.
Bold: heavier weight of type than normal.
Bullet point: a large black dot • used as display feature to highlight *particular* text matter.
Burnisher: a tool for burnishing, made of some hard, smooth substance, with a rounded surface.

C

Calendering: a process producing an exceptionally smooth paper surface.
Captions: descriptive text for illustration.
Centerpiece: the chief object of attention.
Centered text: lines of text aligned centrally to the page.
Clauses: short groups of words forming a grammatical unity in themselves, but which are not necessarily complete, logical, or grammatical sentences.
Collaboration: working together with others.
Composition: the way things are put together.
Contrast: in a photograph or an illustration, the range of tonal variation.
Cross-reference: instruction to the reader to refer to another part of the text for related information.
Curves: templates composed of several changing and graceful curves that may be used to assist drawing.

D

Data: information from which inferences can be drawn.
Design: the underlying plan, or organization, of parts in relation to the whole.
Dropped capital: first letter of a paragraph or section sometimes displayed so it is larger than text, and that occupies the depth of several lines.
Dry transfer sheets: clear plastic sheets on which a typeface is printed in reverse on the back, in a manner that enables the letters to be applied to artwork.

F

Flexible curves: a device in the form of a strip constructed around a lead core that can be bent repeatedly to any desired shape to aid in drawing.
Flow chart: a chart showing the sequence of events involved in a process or activity.
Flush-left/-right: lines of text aligned at the left/or right margin.
Font: typeface containing the complete alphabet, numerals, and punctuation in various sizes, weights, uppercase and lower-case.

Footnote: notes at the foot of the page, separated from the main text but still contained within the type area (usually a smaller type size than the main text).
Format: a term covering the size and the shape of the poster.

G

Gannt chart: a bar chart that displays planning, scheduling, and interrelationships simultaneously.
Graphic elements: parts of which the poster is made up that consist of illustrative material.
Grid lines: a unifying geometrical system selected by the poster designer that governs the arrangement of text and illustrations on the display.

H

Hard copy: a printed rather than electronic version.
Hatching: the process of marking shade by fine parallel or cross lines.
Headings: the titles introducing subdivisions of the text.
Hierarchy of styles: different textual styles to indicate the order of ranking of divisions in the text.
Hot pressed paper: a type of paper that has a particularly smooth surface that is excellent for drawing on using pen and ink.
Hyphenation: a method of breaking and joining words, using hyphens. It is particularly important for word breaks at the ends of lines.

I

Indent: the start of a line is indented (moved to the right) so it is shorter than other lines.
Instant lettering: sheets of dry transfer or self-adhesive letters that may be applied to displays.

J

Jargon: a set of highly technical terms used by members of a particular profession or trade.
Justified text: lines of type that range on two verticals and are of the same length, achieved by varying the space between words.

K

Kerning: adjusting the space between individual letters to achieve a closer fit.

L

Laid paper: paper that shows parallel wire marks, due to its manufacture on a mold in which wires are laid side-by-side.
Laser printer: a device using laser technology to generate text and image reproduction.
Layout: a poster designer's plan, indicating the type to be used, the type area, the position of diagrams, and so forth.
Leading: the space inserted between lines of type to improve visual appearance and legibility.
Legends: captions for illustrations.
Line artwork: illustrations consisting solely of solid black or white without tones.
Logo: a special device for a company name or a trademark.
Lower case: small letters, that is, not capitals.

M

Measure: the length of a line of type.

P

Pantographs: a device for copying a drawing to a larger or smaller scale.
Pasting-up: the process in which the various elements of the poster are pasted to the backing material.
Perspective grids: a device that is marked with a detailed grid that follows the lines of perspective, which may be placed beneath semi-transparent paper to assist drawing.
Pie chart: a circular chart split into slices that show how parts relate to the whole.
Point: a unit of typographic measurement (72 points = 1 inch).
Poster board: the display board on which posters are mounted.
Poster presentation: the process of introducing and conveying the information present on the poster to those interested.

Presentational unity: consistent use of selected textual and illustrative styles among poster elements.

Preserved specimens: examples of an organism (usually) that have been treated to keep from spoiling.

Proofread: checking a version of text or illustrations for errors or omissions.

Protractors: devices used for measuring or constructing angles.

R

Ragged-right (-left): lines of text not driven out artificially (by varying space between words) to the same length based on a common right- (-left) hand vertical.

Raised capital: oversized initial capital that extends above the paragraph it introduces.

Raw data: information from which nothing has been done to extract any meaning.

Reading distance: that distance from which text can be easily read.

Ream: a term denoting a quantity of approximately 500 sheets of paper.

Redundancy: text that exceeds what is necessary or useful.

Rules: black lines of varying thickness, used for horizontal or vertical division and emphasis.

S

Sans serif: typeface without serifs.

Scale drawing: a drawing proportionate to the dimensions of the final design.

Scatter charts: an illustration that displays the distribution of data through time.

Self-adhesive film: an alternative to the dry-transfer process of applying letters to artwork in which the letters are bonded to a waxy backing sheet from which they are peeled before application.

Serifs: "tails" on the ends of characters in typefaces other than sans serif faces.

Software: a set of instructions governing the operation of a computer.

Straightedge: a reinforced or specially fitted edge that ensures an accurate and unwavering line.

Subheading: a heading for an internal division of a piece of text.

Superior letters/figures: small letters/figures set on the shoulder of the main type, so they are above it.

Symbols: graphic characters.

T

Take-Home message: the key piece of information or opinion that results from your research.

Target meeting: the scientific meeting toward which the poster display has been focused.

Tautology: repetition of the same idea in a sentence.

Templates: shapes, usually in the form of thin plastic sheet that you draw around to assist in drawing.

Textual elements: parts of which the poster is constructed that consist of reading material.

Triangles: drawing aids for rapid and accurate drawing of angles.

Typeface: the particular alphabet design used for the main narrative setting.

U

Upper case: capital letters.

V

Velcro: a type of fastener consisting of nylon strips covered with tiny filaments formed into hooks, which may be pressed on to certain materials to adhere and can be removed by peeling off.

Visibility: the state of being noticed.

Visual aid: a device such as a chart, diagram, and so forth for aiding the learning process through the sense of sight.

W

Wove paper: a term applied to paper made on an ordinary web in which the wires are woven (used in contradistinction to *laid paper*).

BIBLIOGRAPHY AND FURTHER READING

Davis, M. (1997). *Scientific papers and presentations.* New York: Academic Press.

Epple, A. (1997). *Organizing scientific meetings.* Cambridge, England: Cambridge University Press.

Imhof, E. (1982). *Cartographic relief presentations.* New York: de Gruyter.

Maugh, T. H., II (1974). Poster sessions: A new look at scientific meetings. *Science, 184,* 1361.

McCown, B. H. (1981). Guidelines for the preparation and presentation of posters at scientific meetings. *Hort Science,* 16, 146–147.

Miller, S. (1994). *Experimental design and statistics.* London: Routledge.

O'Connor, M. (1991). *Writing successfully in science.* London: Harper Collins Academic.

Parker, R. C. (1995). *Desktop publishing & design for dummies.* Foster City, CA: IDG Books Worldwide Inc.

Reynolds, L., & Simmonds, D. (1983). *Presentation of data in science.* The Hague: Nijhoff.

Rogers, G. (1986). *Editing for print.* London: Macdonald & Co.

Stephenson, J. (1987). *Graphic design materials and equipment.* Secaucus, NJ: Chartwell Books.

Taylor, J., & Heale, S. (1992). *Editing for desktop publishing.* Hatfield, England: John Taylor Book Ventures.

Woolsey, J. D. (1989). Combating poster fatigue: How to use visual grammar and analysis to effect better visual communications. *Trends Neurosci, 12,* 325–332.

INDEX

9681